BUDDHA AND THE QUANTUM

BUDDHA

~: *and the* :~

QUANTUM

Hearing the Voice of Every Cell

Samuel Avery

SENTIENT PUBLICATIONS

First Sentient Publications edition 2011
Copyright © 2011 by Samuel Avery

A paperback original

Cover design by Kim Johansen, Black Dog Design
Book design by Timm Bryson
Cover art, "The Power of Pi," by Jason D. Padgett

Library of Congress Cataloging-in-Publication Data

Avery, Samuel, 1949-
  Buddha and the quantum : hearing the voice of every cell / Samuel Avery. —
1st Sentient Publications ed.
    p. cm.
  ISBN 978-1-59181-106-0
  1. Physics—Philosophy. 2. Consciousness. I. Title.
  QC6.A875 2010
  153—dc22
                                              2010032053

Printed in the United States of America

10 9 8 7 6 5 4 3 2

SENTIENT PUBLICATIONS
A Limited Liability Company
1113 Spruce Street
Boulder, CO 80302
www.sentientpublications.com

CONTENTS

We will use meditation to analyze consciousness. We will see that multi-cellular consciousness is a composite of cellular consciousness, separable into perceptual and observational realms, and dimensionally structured on the macroscopic level. We proceed from beyond the bounds of science to see what science is.

THE ACHE

*T*hales said all was made of water.

The ground, the trees, the rain, the clouds, the animals and plants: all were one form or another of water. It hardened to take on the texture and color of rock or evaporated to become the wind. It could move, flow, change, and take new shape. Its qualities changed while its substance remained. Distinctions in the world were of form only; there was but one universal thing at the base of all that is.

Anaximenes said it was air. Air cooled and condensed into liquid, then froze into rock. Air was more fundamental even than water. From earth and wood and metal it warmed and rarified into flowing streams, oceans, rivers, wisps of smoke, and mist. Air, lighter and subtler than water, was the true primordial substance of which all things were made. To man and the animals it brought life itself.

Heraclitus said it was change: flux was the ultimate reality. You could never touch the same thing twice. The real primordial substance was fire. It flickered this way now, then became that. Solid objects—even water and air—were but temporarily frozen flames of fire. "All things are an interchange for fire, and fire for all things, just like goods for gold and gold for goods." Energy was the base of matter.

The debate was new. It searched for natural explanations, for understandings based on things that people could see and touch. It looked to the world—this world—for answers. There were no supernatural powers to take up where logic left off. Imagination was free to roam the pastures of everyday experience, but what was said had to look like what was seen and heard and felt. This was the birth of Western thought.

Buddha said life was suffering.

Three thousand miles to the east he was saying that you could talk all day long about what the world is made of, but what matters is that you will be miserable until you transcend it. The world was illusory, impermanent. What you think of it now, it will not be tomorrow. Suffering came of attachment to concepts of what we think the world is and what we think we are. Fear of what may happen and craving for what we do not have—the difference between what is and what could be—created the ache common to all living beings. Buddha developed a method to break attachment to the unreality of worldly life. The important thing, he said, was not to understand the world, but to understand experience.

Buddha's method was the infusion of awareness into the simplest of everyday experiences: breathing, sitting, moving, thinking. Look at the things you normally do and see what they really are. Do not worry about putting them in logical order or understanding them intellectually; just look at them. Look at them from the inside. Look at yourself looking at them. With practice, the separate existence of *self* will become as unreal as the separate world that it is thought to be *in*. The seer and the seen become seeing. The goal of his method was not realization of a philosophy, but release from the cycle of death and rebirth. The world did not matter; the goal was liberation. Everyone had to do it on his own.

The West looked out to space; the East looked in to mind.

The West looked outward at the world and focused on what could be perceived and verified by others, bypassing what was merely believed, thought, imagined, or felt. The result is a structure of consciousness we call science. Consciousness has become more than what any person experiences. Perceptual information from any one of us, subject to the rigors of the experimental method, becomes fact for all of us. We experience a world collectively that we only glimpse individually. We are a new form of life. Science expands, grows, settles new territory, becomes the world it creates. It sees planets too distant to be felt underfoot and atoms too small to be touched. It is greater than any of us, and greater than the sum of all of us. But though it structures consciousness it does not know consciousness. It does not know what life is.

The Buddha looked at the experience of life and avoided the distractions of matter and space. He looked at life from the inside. The existence of a world outside of consciousness was a question with which he was unconcerned. His method has not grown with the evolution of human civilization. It stands now as it did when he lived. It does not need science and science does not need it.

Or so it would seem.

Science in the twentieth century reached a point where it could no longer ignore consciousness. At extremes of distance, size, time, velocity, mass, and temperature, things were happening that depended on someone experiencing them. Science discovered something that it did not understand and has not understood since: *the quantum.* It found the quantum in the objective world—in space—but it was not really in space, it *was* space.

Science sees things in space but cannot see what space is. The Buddha's method can. It doesn't, but it can.

I present this as fact though I mean it as a question.

I address it to you.

As I am neither a physicist nor a Buddhist, you should read the following not as instruction but as a personal understanding of modern physics and Vipassana meditation. You do not have to study physics to understand what I say, nor do you have to practice meditation. If you do know physics, or meditate, you may recognize some of what is said.

I might have written *I* or *me* instead of *you,* but it would then have appeared to you as happening in another self.

Bear with me.

BUDDHA

*B*uddhist meditation begins with breathing. Buddhism begins with morality.

Breathing is always there, meditation or no, and that is why it is the beginning. There is no departure from not meditating— the difference is paying attention to it. It is the most commonplace of sensations and there is no reason to consider it, but you pay attention to it. You don't think about it; you look at it. You look at breathing in and breathing out and see it from the inside. You feel it in the rise and fall of the diaphragm, in the expansion of the chest, passing through the windpipe, and at the opening of the nostrils. You do not envision it—it is already there. You do not try to change it. You do not try to make anything happen. You don't do anything at all; you just look.

But you do not look for long. You try to keep your mind on the breathing, but it wanders. It passes on to other things. Soon you are thinking of what you have to do later today, of what someone said last night, of the discomfort in your lower back and how you would like to move. Soon you have forgotten the breathing. You think of what you might have said, develop a plan, straighten and move to a new position. Your mind wanders to the shopping list, to what you will wear, and you remember that you are supposed to be meditating. You bring your mind back to the breathing—the rise and fall—the air passing through your nose. You pick up where you left off. You are back where you were and you watch for a while: in... out... in... Your mind settles, but soon wanders again. It passes on to something else, and as it passes you watch how it passes. You watch how the breathing becomes something else. You look at the difference between breathing and thought. As looking becomes thinking, you bring your attention back once again to the breathing, to looking at the breathing. Always look at what the mind does, and bring it back. In this way you learn concentration.

There is nothing mystical or spiritual about concentration. Everyone learns it in one manner or another. A musician learns to listen to pitch, a carpenter to saw a straight line, and a pickpocket to watch how people move before he strikes. These all require concentration. They are ways of focusing the mind to ensure that doing is done well. They are ways of applying what is thought to what is done.

Concentration in meditation is without doing. It is the application of the mind to what is seen or felt or thought without changing anything. It is creating awareness without an agenda.

If there is a purpose in mind, concentration can be helpful or harmful depending on the purpose. Concentration of the mind that one learns through meditation can be put to good purposes or to bad. Horse thieves, bank robbers, and assassins can all improve their skills through meditation. That is why, though Buddhist meditation begins with breathing, Buddhism does not begin with meditation. It begins with morality. It begins with putting oneself in proper relation to others. In that way meditation will lead not to enhancement of the self but to its transcendence.

Compassion for all living beings—human and non-human—is at the center of the Buddhist religion. It is what makes Buddhism a religion. When I see others suffer I experience it as my own suffering. As I would not harm myself, so I will not harm them. I will do unto others as I would have them do unto me. I do not feel pain in the bodies of others as I feel it in my own, yet their pain is as real as my own. I choose to accept it as such; I choose compassion. In this way I will understand being. I will transcend the physical separation between others and myself, and I will see that being is larger than self.

There is no logic to it. There is no mechanism that ensures I will be better off treating others as myself. There is no proof that other people or animals feel anything at all. There is no arithmetic that brings to me the amount I take to you. There is only the intuitive reality of existence beyond direct perception. There is only a guess that consciousness is greater than me. This is the essence of all religion.

It is the essence of civilization.

And now, it is entangled in physics. If consciousness is self, we will not understand what we see. We will understand neither relativity theory nor quantum mechanics. The role of the observer—consciousness—is at the heart of understanding modern physics; we now know through science itself that the world is the way it is because someone is looking at it. But who? You?

Me? A scientist? A dog? What is the relation between one consciousness and another? How may consciousness be other than
self? Science has never before concerned itself with such questions. Now it must, because of what it has discovered.

Breathing is partly under the control of self and partly not. You
can decide how to breathe if you want to, but you don't have to.
It will happen if you decide nothing. You can hold your breath
and watch what the self thinks and does to get it back, or you
can watch the breath without trying to change anything about
it. The object is not to defeat the self but to watch it, and to become aware of how it curves consciousness.

   Concentration must be learned to see clearly on the inner
path. Practice begins with small sensations somewhere in the
body: in the diaphragm, the nose, or the skin above the upper
lip just outside the nostrils. If you look there closely you will see
what is happening. You will see reality at its most elemental
stage. There will be other things going on beside your object of
concentration; you take note of them without allowing your
mind to be distracted. They are as real and important as what
you are concentrating on and you should not ignore them. Just
let them pass. Don't try to tune out the rest of the body or the
rest of the world; just watch it fall away as you focus on the area
of concentration. Look deeply into this one place. Catch the
mind as it wanders, and bring it back.

   Once you have developed a degree of concentration you may
move on to something else. You look at the entire body from
the inside. Start at the top of the head and move slowly through
the neck, shoulders, arms and hands, chest, abdomen, hips, legs,
and feet, and then move back up. You try to see and experience

everything exactly as it is. You feel air on the skin, hairs tingling on the back of your neck, bones rubbing in joints, the heart beating softly in the ribcage, digestion in the stomach, throat, and intestines, the weight of your body on the pelvis. Yet you do not think of it this way. You do not envision bone against a bone or a hair waving in the air. You look at the sensation, not at the object of sensation. You look at what is. It is there in the moment and you do not have to think of it in any particular way. You do not have to envision or describe it. You do not have to think of it at all—just look at it. The mind tends to picture what is happening as if looking at it with the eyes, but you are not seeing it visually. You are seeing it from within. You are looking at kinesthetic and chemical sensations that may not have the "shape" of visual sensations. Look at them exactly as they are. Do not try to envision them in space. Look at what you are actually experiencing without putting it into a context. The mind will always want to take what you experience and put it into categories and words and shapes that make sense to you and that you can tell others about, but you are not doing that. Just watch the experience as it happens and let it go. Forget about describing the experience. Forget about remembering anything. You are trying in meditation to go beyond description to the thing itself.

You are trying, but it does not work. You are looking only at the sensation, and you cannot quite get at it. You cannot quite touch it. You reach out with your mind to catch it but it will not stay still. It is always out of reach. The thing itself is not exactly there. The sensation is real, but what is causing it? What is "out there" behind the sensation?

The Buddha said that beyond sensation there is only *emptiness* and *impermanence*. Sensations, as elaborate as they may seem, are without content. They have no substance behind them. They pop up and stay for a while, then disappear, no matter how substantial they may seem at the time. They arise and go away.

The mind puts sensations together. It shapes them into thoughts and concepts and fits them into categories, relating them to other sensations, and thereby comes to know something *about* them, without knowing the thing itself. The thing itself is beyond mind and that is why there is no way to understand it. "Thingness" is itself a concept—a creation of the mind. Thingness cannot be anywhere that there is no mind, which is where you are trying to go.

It is for this reason that meditation does not sit well with the mind. It is not understandable—you cannot say what good it is or what it is for. Its business is to confound the conceptual mind, and the mind will never accept it as a good thing to be doing with your time. The mind will come up with other things you could be doing besides meditation.

Sensation is accompanied by thought, and thought by sensation. The mind and the body are thoroughly entwined. The difference is time. Both are at points in time, but the body is *extended* in time. It is less impermanent.

The Time Dimension

Stop and look at the last paragraph more closely. The difference between thought and sensation—between mind and body—is time. This is may seem a strange thing to say. But look closely: everything you experience is reducible to thought or sensation. Both happen at points in time. Sensations seem to last in time, at least for a while. They constantly change—get stronger or weaker, louder or softer, etc.—and disappear eventually, but the ache you are feeling in your elbow now, or the lamp you are seeing on the table now, is the *same* ache or lamp that you were feeling or seeing a minute ago or a week ago. The ache may have eased a bit, or the lamp moved to a new position, but it is the same ache and the same lamp. Each has an extension in time. Thoughts, however, pop in and out of time. You experience similar thoughts later, but not the same thoughts.

This may seem too precious a distinction, but it is key to understanding the perspective I am trying to develop in this book. From within time, sensations appear to exist as extensions and thoughts as points. But stepping out of time—looking at time from the outside—you get a view of the same thing from a new perspective: time is the coordination of thought and sensation, a sort of pivotal plane that ties sensation and thought together at specific points. You're not sitting in time, waiting for thoughts to happen while you experience hearing, feeling, seeing, etc.: you're seeing thoughts lined up and joined with sensations at certain points. "Stepping out of time," of course, requires detachment, or the removal of conscious experience to a distance from what is experienced. It takes a lot of practice. It is impossible to describe, and impossible to "do" in the normal sense, as there is "nowhen" else to go. But it is through stepping outside of time that one realizes that thought and sensation are of the same primal substance, and that time is what distinguishes them. Time organizes experience. If detachment is accomplished through transcendence of time, it can be accomplished also through transcendence of space, which further organizes experience. Detachment from time and space reveals dimensions as fundamental structures of consciousness.

As you move concentration throughout the body from head to foot, a problem arises. You want to be aware of everything that is going on while maintaining your concentration at the same time. You want to know the entire body as a whole—a singularity—but you want at the same time to be aware of everything about all of its parts. This is a contradiction. If you look only at the details—a digestive feeling in the stomach or an ache in the lower back—you lose the big picture. If you look only at the big picture you lose concentration because you are not looking at anything in particular. Your mind begins wandering again. This is why meditation takes so much practice, and why you have to understand it as practice. You have to go through a lot of things that do not produce results at the time. You have to forget about results entirely.

You are watching skin sensations at the top of the scalp, then around the eyeballs and ears and the back of the head. You feel something pulse within the brain. Then you feel air in your nose and throat, the solidity of teeth and jaws, and moisture at the back of your tongue and down the esophagus. You move to the gentle heaving of the chest, watching the rib cage slowly rise and fall. Your concentration is good, perhaps better than it has been for some time. But as you move on to the rhythm of the heartbeat, a dull pain develops in your hip where the femur joins the pelvis. You have been sitting for a while and this part of your body has had enough. It wants attention. You have been taught to watch how the pain develops—where it moves, the shape it takes, how it disturbs your mind, and how it recedes—to be "objective" about it and not react. Do you stop looking at the heartbeat and look instead at the hip? Do you tell the hip to wait its turn and stay with moving systematically down through the body? Do you try to do both at the same time? The choice itself pulls concentration away from looking at either. Is there a better way, a method that avoids such choices? Is there a technique that glides automatically from one concentration to another without requiring a decision? How is it possible to look past the mind if mind itself is involved in the technique?

A better question is: what or who is it that makes the choice? When a decision must be made to continue concentrating on the heartbeat, or to move over to the hip, or to move the hip to a new position, who decides? The easy answer is *you*. You decide. But what are *you*? The Buddha, if he could look inside, might say, "I see no *you* here. There is a heart beating and an aching hip joint, there are anklebones against the floor and hairs at the top of the scalp, there is air passing in and out, but I see no *you* here." He would say that the *you* is no more than a bundling together of sensations—a string wound around a pile of sticks. It is an illusion that arises only as an *attachment* to the sensations.

As one becomes familiar over the hours and years with the subtle experiences of body sensation one notices that they float by themselves. They have a life of their own that is not you. You learn to avoid thinking of *my* heartbeat, *my* leg, *my* ache. As you look at each sensation without thinking about it, you come to see it is exactly as it is—that is, without an object out in the world causing it, or a self somewhere inside perceiving it. The seeing does not require a seer or a seen. The attachment that is the illusion of self becomes plain. The pain you feel in the hip comes with an attachment that the heartbeat does not have. It wants you to *do* something. It wants you to react. It wants to keep you tied to the endless cycle of death and rebirth. You know that someday, when you are better practiced, you will rise above the attachment and be able to look at the hip and the heartbeat on equal terms. You will be able to look at them without reacting to either. But right now the pain is all you can stand and has spread to your upper leg. So, you move it. That is what *you* are.

Concentration is extremely difficult to control, and you do not want to control it as such. Control empowers self. You want rather to watch whatever your mind is doing and gently shift it back to what it should be doing. Moving it slowly through the body gives your mind a place to go back to when you catch it wandering. When you start thinking of a new route to drive between work and the hardware store that avoids a left turn on Main Street, you consciously bring yourself back to the tension in the knee joint, and then on to the shin and ankle. At the moment you start to wander again, move on to the next place. Hit every spot in the body, whatever is going on there, one place at a time, until you have covered it completely. Go from top to bottom, from side to side, and from front to back. Concentrate on one spot at a time, but do not tune out other sensations in other places. This is a difficult balancing act, but as concentration improves, you will be able to scan larger portions of the body at a time, and to move more quickly without skipping over anything. With exceptionally good concentration you can scan the entire body with a single breath, moving down from head to toe on the exhale and back up on the inhale.

Teachers of the Buddhist method speak of gross and subtle sensations. Gross sensations are *things* that you feel: an arm, the back of the throat, or an itch in the middle of your back. They are the *what* that you look *at*. You are aware of the most powerful of them when you are not meditating. Subtle sensations, on the other hand, are what you feel when there is nothing there. They are everywhere in the body but not specific to anywhere. It is hard to describe them—they are less than a tingle, less than a breath of air, less than a brush or an itch—much less. You might think you are imagining them, but they don't fly around and get caught up in other things. They are more like a background that you don't notice because it is always there. They are points in time, not extended in time. It is the shapes they make that are extended in time. You become aware of them as you scan the body and do not see anything in particular.

Subtle sensations are the point at which sensation arises from thought.

Subtle sensations are known in Buddhism as *kalapas,* a Pali word meaning *smallest things.* Modern Buddhists often define them as *subatomic particles,* though they do not exist objectively. They are units not of matter in the objective sense, but of experience. Only you feel them. You may hear other people talking about them but only you feel them.

Their sensation is often pleasant, and sometimes described as a *flow*. The effect of experiencing them throughout the body can be as blissful as anything imaginable, which causes problems for Buddhist practitioners. The feeling of kalapas flowing through the body can become a goal—a sort of payoff for hours of pain and boredom from sitting still—but once you want it you begin to crave the good feelings when they are not there. This is the opposite of what meditation is about. You should be looking at what *is* there: blissful, painful, or neither. If there is bliss, fine, go ahead and feel it, but you should not try to make it happen. You should not try to make anything happen, or try to feel something that somebody else told you about. If you do, you will wonder if all the pain you have to go through is "worth" the pleasure that you think is owed to you. Meditation is not about you, and the pleasure you may feel from time to time is not yours any more than the ache.

The blissful feeling of kalapas may make you not want to go any further. You may think you have reached the goal of meditation when you have reached only a milestone along the way. If you become attached to the good feeling you will begin to crave it. You will wonder why it does not arise automatically every time you meditate, and you will become frustrated. Detachment from subtle sensations is as important as detachment from pain, boredom, or anxiety.

Gross sensations do not disappear altogether when you begin to feel subtle sensations. If your concentration is not good, you will see only gross sensations as you look through the body. You will be aware of the diaphragm, the heartbeat, or pain, but if the mind is sharp, gross sensations relax and fall away as kalapas begin to make their appearance. You become aware of them once there is nothing tugging at you to scratch, itch, or move a leg. They are tingling, scintillating sensations that you have felt many times before but that you considered unimportant. They were just a glowing feeling you experienced here or there that you soon forgot. You did not pay attention to them because you did not have to: they did not require anything of you. Unlike gross sensations, they did not need you to *do* anything so you passed over them. But you see them now because you have practiced concentration. You see them because you have learned to look at what is there, whether or not it seems useful or im-portant.

Gradually, with practice, you feel them everywhere and the solidity of the body begins to dissolve. This is the state of *bhanga*. There is no sense of *my body*. With good concentration you can sit and watch shapes and sensations come and go—ap-pear, swell, burst, and pass, without dragging you with them.

The words *kalapa* and *bhanga* have more specific definitions than I give them here. I lend them more importance in this dis-cussion than they have generally in the Buddhist tradition. My understanding of what they mean is derived less from study of Buddhism than from direct experience. Precision in my use of them suffers accordingly. They are what you see, or feel, when the mind and body are still. Bhanga is like the static back-ground of a television screen without a broadcast signal; the kalapas of which it consists are like electrons popping in and out of existence. You can almost hear them. From an objective point of view kalapas are, in fact, electrons jumping across synapses from one nerve cell to another. But they exist only as tiny units of experience; they are little pops and crackles at the very foundation of perceptual consciousness.

Using words of any language to describe meditation is like playing the piano with your knuckles. Any music that results is in the ear of the beholder.

Kalapas are not objective experience and are not located in space. Nobody else sees them. They seem to arise from various parts of the body, but as you experience them, they lose their association with any specific location. You may come to see that the reverse is true: parts of the body arise from *them*. As you look into the body, a sort of space arises with which to associate gross sensations: a right and left, an up and down, a height and width, even a depth. But it is not the same space as outside the body. It is not as clear, not as distinct or delineated, and it is not measurable. You can tell more or less where a left arm or a right foot is, but you cannot tell as precisely where they are as when you see them. The body is a primitive sort of space that is spatial only in the sense that where you feel something with your eyes shut is where you will see it when they are open. Yet it is where the sense of space outside the body comes from. Visual space is an outward projection of the space that you feel in the body. It is why what you feel is felt *inside*.

Gross sensations are like objects on a computer or movie screen; they have form but no underlying substance. They consist of nothing beyond the sensation itself. As you become detached from them you watch them from an objective stand-point. The "body" you experience is not your body. It is not you at all. It is a ground of universal experience that you access through meditation. The mind does not grasp this easily because it is used to thinking of each person with a body separate in space, and of each body with its own consciousness. How can experience within the body become universal? Things that are entirely outside the body do not produce sensations within it. Sensations within it are not experienced in other bodies. How does experience entirely within the body become that which everyone experiences?

The concepts of *inside* and *outside*, of *within* and *without*, are derived, like many concepts, from visual experiences, from space outside the body. The concept of one thing being separate from another is derived from spatial experience. But the body is not spatial in the way visual perception is spatial. Portions of the body seem to be in different places, but none is entirely separate from another. Detached from the self, and from any sense of doing, bhanga is not separate from other people, from animals and plants, or even from inanimate objects. It is what everything feels. Gross sensations tend to form into objects, and thereby require a subjective *you* to respond to them, but the ground of being experienced in bhanga is common to all things. It has no subject or object.

The whole world is right there.

Bhanga is a state of perfect detachment. It is a transcendence of the body, of the objective world, of space, and of self. You see everything at one timeless moment. There is nothing that is not there, and there is nothing else that you need. Space is not there and you do not need it. With eyes closed there is no space.

Some will suggest that bhanga is a state of mind. It is no more than a good, relaxed feeling one gets from quieting anxieties, and there is nothing universal about it. It is just the way things seem to you. They are, of course, absolutely right. Any concept of bhanga, whatever one has heard or thought about it, is purely mental and not universal. The word and the thought behind the word are not the experience. If you have had the experience your memory of it is not what it is.

The universal experience reached through meditation is the *universe experienced*. It transcends the separation of mind and body. It is all there is; there is nothing beyond it but confusion, craving, and ignorance, and these boil away in time. This is not the objective universe of science—not the universe of space and time we experience through visual consciousness, and does not claim to be. Buddhism does not deny the existence of the objective, visual world; it denies only its importance. It may or may not be there; arguing one way or another does not gain release from it. The point of Buddhism is not to come up with answers to life's questions, but to be released from life, to attain personal liberation. The direct route avoids entanglements and distractions that the world presents.

"Do not worry about metaphysical questions," the Buddha might say. "Do not bother with philosophies. Work on detaching yourself from what you experience. Work on your meditation." The Buddha suggested once that if walking through the forest one day a man suddenly found himself shot in the back with an arrow, he should not ask what kind of arrow it is, where it came from, and who might have shot it—he should pull the arrow out! Life is suffering; the goal is to escape attachment to it. One should not be concerned with intellectual games. Leave speculations behind.

A way to avoid distractions is to keep your eyes closed while meditating. There is plenty to worry about already, with all the aches and pains in the body and the relentless churning of the mind—plenty to distract you without adding objects and motions in the world around you. Shut down as much as you can. You cannot shut it all down. You cannot turn off the body or shut down the mind—these you must work to transcend. But you can close your eyes. You can shut down this much needless distraction. Questions will linger without answers, but questions are only playthings of the mind.

Pull the arrow out!

Buddha's method does not need books like this. It does not need anything that has happened since the Buddha himself walked the earth. The method is not only universal, it is eternal: it is now as whole and complete as it was twenty-five centuries ago. It has survived so long and continues to this day because of its consummate simplicity: life is suffering; here is the way to escape!

And so, we leave Buddhism at this point. We will not speak of *nirvana* or higher stages of meditation. We leave with an intellectual understanding of bhanga, if not its attainment. It is made of kalapas, tiny currents of subtle sensation that we are aware of only with a quiet mind. Kalapas, though we are aware of them through the body, are a transcendence of self.

We leave Buddhism now because of what has happened in the world since the time of the Buddha. Much of the outside world has found its way in, and become what we are. We are no longer what we were then. Salvation is no longer personal. We have seen with our eyes the arrow that is in our back, and would like a better look.

The Buddha did not know of the quantum, and could not have been looking for a way to explain it. The quantum came looking for him.

WHAT IS THERE?

What is there?

We stray from Buddhism because of science. Science is much of what we are, and it has happened since Buddhism.

We stray now also from beliefs, creeds, ideologies, and schools of thought, whatever they may be. We set them aside, for the moment. We do not forsake them; they will there when we come back. We only step around them, for the time being, noting where they are and how to get back. We step carefully around what we have heard or read or thought before. We will look at what is, with a quiet mind.

What is really there? Perhaps you do not meditate regularly, or at all. You nonetheless see what is there. Meditation will help you see it clearly, but it will be there if you do not meditate. There is no reason not to look.

What you see are thoughts and images, feelings, disturbances, and sensations flying about—pieces of things you have seen or felt or thought before. Close your eyes and look at them more carefully. Some are clear and sharp-edged, others vague and indefinable. You feel the chair against your back, think of what you will have for lunch, watch a fleeting thought of something you should have said, hear the cat at the windowsill, remember a friend you haven't written, wish that it would stop raining, sense a discomfort in your lower leg, hear the bark of your neighbor's dog, and think a piece of a thought about a problem you are having. Each sensation, each image, rises from nowhere, comes to view for a split second, and is gone.

But what is really there? What is there of the world in this? What of you? So much is thought—mere imagination— what is *real*? What is hard, physical, concrete? Is reality other than what you are seeing—something you do not see—causing you to see? Are the thoughts and images you see as real as the chair against your back and the dog in the yard next door? Is it all just in you? Do you only think there are a chair and a dog?

Now you are thinking about thinking. Thoughts are piling on top of thoughts and there is no way to straighten them. What are thoughts? You are poking a stick to see what things are, but you can't poke the stick at the stick. You can't think what thoughts are because that is what they are.

But you can watch them.

You notice that they take any shape. You can lump them together this way or that into bigger thoughts, or tear them apart into smaller thoughts. You can change the big thoughts to fit more small thoughts, or change the small thoughts so that more of them fit into big thoughts. If they do not fit as you would like, you push them until they do. You arrange them almost any way you please. You can say, if you choose, that *real* means a thing—a substance—that you do not experience directly but that causes you to think, hear, feel, remember, hope, see, etc., i.e., that *matter* is what is real. Or you could say that the only real thing is *you experiencing* all the images, sensations, feelings, thoughts, aches, pains, joys, sounds, smells, memories, aspirations, etc.—that the only real thing is *you* because that is all you actually experience. Or you could question the *you*, by defining *experience itself* as the ultimate real, saying that *you* are a construction of that more primal reality. There are many ways to think.

Some are better than others. Some big thoughts fit more small thoughts without as much pushing. This is important, because we need good thoughts to do good things and be good people. We need good thoughts to live our lives and to secure things needed to keep living. But thoughts, good or bad, are fluid. They change. A thought is good, useful, inspirational, essential—it may recur over a millennium or two—but it is not final. It is only what you think.

Watching thought is more fundamental than thought.

Nothing can be said about the watching—about consciousness. There is nothing more fundamental of which it is made. It is not part of anything. Everything is made of it. There is no way to say what it is because there is nothing it is not. At the bottom of all things is watching.

Watching is what is there.

Thought is not everything because I can think about a thought. I can get outside of it. A thought may be inseparable from the whole but it is never the whole whole. It is always a part. There is no outside to watching a thought, but there is an outside to the thought. There is always something it is not. Where or how thinking proceeds from the watching I do not demand to know because *where* and *how* do not yet exist in the manner of thinking I have chosen. They come later. Asking where thought comes from is like asking where space comes from. There is no *where* without it.

Thought is a word I use for what everything is made of. It is not the right word exactly because it has other meanings. Other people might use another word or say that *thought* is something else; I choose it because I need a word, and *thought* is closest. Some think that *thought* is all that is *not* real. That way of thinking works, but then we're back to defining *real*, and I don't want to get caught in a tangle of words before we get started. I have used words like *image, idea, concept,* etc. to mean the same thing as thought, but there are other meanings for these words, too. Each will work and not work. I don't really mean a word at all—what I mean is a moment of experience. It is the thing itself that I am getting at, whatever the right word may be. The word can only be a symbol that stands for but does not contain it. If you look closely you will see the thing itself and know what the word stands for. Don't just read this.

Everything is made of thought. Thought and perception are of the same primal substance. (I use the words *perception* and *sensation* more or less interchangeably.) Thought is either loosely configured as *conceptual* consciousness, or it is dimensionally structured to become *perceptual* consciousness. A perception is a thought in a dimension. (*Observational* consciousness is a further construction of perceptual consciousness that I will get to later.) Everything is one form or another of thought. Dimensionally structured thought— experience in space, time, and mass—is the physical world.

I don't know that this is true. I only think so. And I think the thinking is more fundamental than the *I*. I think even the *I* is made of thought.

Thought is more fundamental than self. There is nothing outside of thinking doing the thinking. This turns out to be a good thought because smaller thoughts—things like the physical world and self—fit inside it, without too much pushing. They fit with less pushing than any other way I can think of.

Thought is more fundamental than *I think*. Thought, and not self, makes us feel alive.

Thought, therefore being.

This takes us back to a path we took nearly four hundred years ago, at the beginning of modern science. We thought then that science came from a common sense universe, an external world made of space and time before there was thought. Inside of this universe were objects made of matter, some of them alive. Consciousness was a parallel reality that impinged on living bodies in mysterious and enigmatic ways that no one understood then or has understood since. We thought that thought happened inside of self. René Descartes showed us this earlier path and we took it. It has served us well for hundreds of years. But now, since the quantum, it leads nowhere. Descartes said "I think" instead of "thought." He reduced experience to its barest essential, *thought*, but *self* snuck in when he was not watching. He did not doubt enough. He did not doubt the existence of self. This is the difference between Eastern and Western thought. The path he established leads through science and the quantum to solipsism, to the world inside of self. Thought inside of self is the world inside of self. To understand the quantum, we must go back to where we were four hundred years ago and try another path.

Thought, therefore being.

The path I have taken begins, instead, with watching. It watches thought create itself and become everything else, including self. Of many there, I choose this path because it is clear to see, and because there must be a clear path to science. This is of extreme practical importance because science will have its place in our lives, one way or another. It will not stay where it is. Science will have a new place, and the place it will have will be the solipsism if we do not find a better place.

Science has undermined the metaphysical foundation we built beneath it. Without attempting to do so, it has seen beyond the space, time, and matter that we put it on top of. It has seen conscious experience beyond the dimensions, and does not know what to do with it. Science no longer knows what it is looking at. This is disturbing because we thought science was everything. We thought it was all there was. We thought it could tell us what everything was made of. But it cannot. It does not know what consciousness is, and cannot know. It has tried to deny the importance of consciousness, or to place it inside of itself, but it does not fit. We suspect, therefore, that consciousness is not to be found in science—that science is to be found somewhere in consciousness. Science is likely to be a special structure of consciousness. To know, we must look back to what is really there. We must look outside of science to see what it is.

Science concerns itself with neither conceptual nor perceptual consciousness, only with *observational* consciousness. It is not interested in what you think or what you see or touch or taste, only in what anyone—any *observer*—says he sees or touches or tastes, under the same conditions. Things are true in science not because you see them, but because everyone sees them, or says they see them. To be science, it has to be observable, or *potentially perceivable*, by anyone. This is a realm of consciousness beyond perception. It is not in me or in you. But it is a realm of consciousness built from perceptual experience. What we all see is a special structure of what you see.

Observational consciousness is *potential* perceptual consciousness

The systematic creation of observational consciousness is science: what everybody can see.

Here is an outline of where we are going:

We begin with watching thought create thought. (A word for this, of course, is God.)

Thought creates *self*.

Thoughts that arise within time and space become perception. Five realms of perceptual consciousness emerge.

Two of them—the chemical and tactile realms (taste and touch)—correspond with time dimensions and the other three—the olfactory, auditory, and visual realms—correspond with space dimensions. (This will take some explaining.) We will show that each dimension is a *potential* for the sensory information of its particular realm, and that the physical universe (space-time plus mass) is an inter-coordination of five sensory potentials.

This is why you touch something *where and when* you see it. This is why you think it is made of material substance. Try it.

All five perceptual realms are dimensionally coordinated with a sixth realm, the observational realm. That is why I say I see something *where and when* you see it. You don't see me seeing it; you hear me *say* I see it. Observational consciousness is the dimensional inter-coordination of what we all say we see.

The physical universe is not an external structure that we are in. It is the structure of consciousness. There is no existence outside of consciousness.

Unicellular consciousness is limited to the temporal, or chemo-tactile realms. Single cells experience only chemical and tactile sensations; they do not smell, hear or see. They do not experience space. Space is a creation of multi-cellular consciousness. The olfactory, auditory, and visual realms are the wholeness of multicellular consciousness, but they are also *reducible* to the chemical or tactile experience of single cells. They have to be. That is why the physical world is not perfectly continuous as seen through multicellular consciousness. That is why weird things happen in physics at the most fundamental levels.

We will show that observational consciousness is a dimension-like structure of *order* associated with the behavior of living organisms. Unlike inanimate objects, observers move in an orderly manner that tells us they are alive. Observational consciousness is built from the experience of individual observers the way perceptual consciousness is built from the experience of individual cells.

Then we will say something about what *you* are.

The path we follow is a winding path, to be sure. It leads through densities of verbiage that block the view from time to time. Meditation and quantum physics are hard to talk about; they defy normal tools of conceptual thinking and lend themselves to everyday language only grudgingly. To capture them entirely in words is not possible and I will not attempt it. Instead, I will point at what I have seen, and hope that you can see it as well. It will not be in the words, but in what you see.

The path begins with what you experience directly and leads through what you know or may have heard indirectly about the physical world. You can stop anywhere along the way and know where you are.

It begins with looking at what is.

As you look, you notice that some of what you experience differs from ordinary thought. It seems to have something solid behind it, something substantial. When you hear a bark, you think of more than the bark: you think *dog*. When you feel a sensation at your back, you think *chair*. There is no actual experience of *dog* or *chair*—only the bark and the sensation at your back—but the perceptions make you think there is. How do these differ from ordinary thought?

The difference is in how they are put together. Ordinary thoughts are put together with other thoughts to form images, plans, schemes, memories, dreams, fantasies, etc. that are as *real* as perceptions, but that lack the structure that makes them seem substantial. Perceptions—sensations that you hear or feel, smell, taste, or see—are put together in a *dimensional* context. That is the difference. They are of the same primal substance as thought but come in space and time dimensions. It is their *coordination* in space and time—their dimensional relation to one another—that makes them seem substantial. You see things *where* and *when* you touch or hear them. That is what makes them seem material.

Let us see if we can find the point at which this happens.

Look at what is there, with your eyes closed. Let us suppose that you have plugged your ears and nose, so that you do not, for now, hear the bark or smell the scent of lavender in the air. What is the difference between the thoughts you see rippling across consciousness and what you feel in your body?

Thoughts flicker. They wave back and forth for a split second, but do not stay put for long. It is hard to find them once they are gone. If you do find them they are not the same. They occur at points in time but they do not endure through time. What you feel in your body, on the other hand, stays around for a while. It changes, and eventually goes away, but it is extended in time. You can look at the feeling of your foot against the floor and then, a minute later, look at it again. This makes it feel more real, more substantial than a thought. There is still no experience of floor—your eyes are closed and you don't see anything there—but the time dimension highlights the sensation. It puts the sensation in a different category: this is not something you are just thinking.

You can see this easily without formal meditation. Without spending hours sitting quietly with your legs crossed you can see the difference between thought and tactile perception. What is more difficult to see is that the things you feel in your body are gross sensations. They are made up of much smaller, finer, harder to recognize subtle sensations that you are unlikely to see until you look for a long time. Seeing in any situation takes concentration and familiarity with the territory. If you step into a forest for a few minutes you will see, hear, and smell all the wonders of nature, but you will not be able to tell one tree or bug or flower from another. If, on the other hand, you spend hours or days or years in the forest you will see all the wonders and know them one from the other. Similarly, if you take a short glance at the body you will see very little of interest. If, however, you spend an hour or two looking at it carefully, you will come to see the tiny subtle sensations called kalapas. You will feel every cell in your body. You will realize that what you call an arm or neck or a finger is a range of subtle cellular sensations that you experience in aggregate form. It is something that your nervous system has collected and channeled into your brain. When you are meditating you can see this plainly. When you are not meditating your cells have to shout—and shout together—to get your attention. You can't hear them individually.

Subtle sensations are the tiny points in the body that come together when there is a feeling in one part of the body or another. They are hard to see because they are so small and because there is no empty space in the body. There is no "space" in the body between one gross sensation and another. Body sensations do not have crisp edges like visual images. They blend into one another: it is hard to say where one leaves off and another begins. Space is not as clear and precise in the tactile realm as in other perceptual realms; its dimensionality is temporal.

Kalapas, like thoughts, are points in time. They are not objects.

Gross sensations are extensions in time.

Kalapas are the freezing point of thought and the boiling point of sensation.

Kalapas are less what you actually feel than the background against which you feel it. They are the empty template of sensation—the points at which thoughts enter time and become sensation.

The difference between body and thought is time.

The body is always there.

Watching, attempting to see thought, you see time.

You have to look at kalapas to have any idea what I am talking about.

Subtle sensations flicker out of thought and into time as waves ripple and collapse into particles. Thought becomes body.

A kalapa is hearing the voice of a cell.

PLANCK

The Earth, Water, Air, and Fire of the ancient world became the solid, liquid, gas, and energy of modern science.

The Western mind, fascinated with what could be seen and touched, concentrated on knowing the world outside of mind, on what could be proven to be true and known to all. The objective world was thought to exist independently of conscious experience, to be there whether or not anyone was looking.

Consciousness is not well understood in the West. It is assumed to exist inside of space.

In the years leading up to the twentieth century the general picture of the objective world was becoming clear. Scientists were feeling confident about what things were made of and how they worked. Matter was made of atoms, planets stayed in their orbits by the force of gravity, sound was waves of air molecules and light was waves of ether. The Periodic Table of the Elements became the backbone of modern chemistry, Darwin's Theory of Evolution of modern biology, and stratigraphic succession of modern geology. In physics, heat was no longer thought a substance, but the motion of atoms and molecules, and thus a form of energy. Experiments with electricity and magnetism led to a unified theory of electrodynamics. Light was discovered to be a form of electromagnetic radiation, other forms of which were the newly discovered radio waves and x-rays.

The world was rational and knowable. All was not known, but could be known in time. What held the world together and made everything work properly were conservation laws. The law of the conservation of mass governed chemical reactions—total mass before any chemical change, such as combustion, was equal to total mass afterward, even if some mass was hidden in escaping carbon dioxide. The law of the conservation of momentum governed elastic collisions between physical objects: if you knew the mass and velocity of any two colliding objects, and the angle at which they hit, you could tell exactly what their directions and velocities would be after the impact. The law of the conservation of energy covered all chemical and physical interactions, including non-elastic collisions. Energy of motion was convertible into potential energy, as when an object is thrown up vertically and slows against the pull of gravity, or convertible into heat when friction slows the motion of two impacting objects. It was believed that if one could know the mass, the position, and the motion of every object in the universe at any moment in time, it would be possible, through the laws of physics, to calculate the mass, position, and motion of the same objects at any other time—past or future. Everything happened according to natural law. This was the age of classical physics. It was perfect, practical, beautiful, elegant, orderly, and complete, and there was no reason to believe it would ever be otherwise.

The world of classical physics is the world of common sense.
It is where most of us still live.
It does not exist.

Beneath the physics of the nineteenth century was a metaphysics of which few were conscious. Most scientists would say that the work they did had no metaphysical basis at all. It did not need a philosophy. It rested on proven fact and there were no assumptions behind it. Space, time, and matter were so obvious, so apparent, and so clearly constituted the structure of the universe, that it was useless to think about them. They just were. There was no reason to suppose otherwise—unless, of course, you were a philosopher and liked to quibble. Space was continuous, absolute, and infinite in all directions. Every object occupied a distinct and unique position in space. If the object were in motion, it passed through a series of distinct and unique points in space. Motion was absolute in relation to a fixed universe.

A problem that classical physics did face was that it could not define absolute space or determine where a fixed point of reference lay. Was the earth at rest, or was the sun at rest, or the fixed stars? What universal frame were things moving in relation to? Nobody knew. A clue lay in the *ether* that filled all of space and provided a medium for the transmission of light waves. This medium had to be at rest and would therefore provide a fixed reference for all motions in the universe. Experiments were performed in the late 1800s to detect the earth's motion through the ether and it was assumed that the problem would be resolved before long.

Space was thought to be infinitely divisible. No matter how many pieces you cut it into, each piece of space could be infinitely divided into yet smaller pieces. Consequently, there was no limit to the precision of measurement; bigger and better microscopes would produce better and more accurate images. More decimal points could always be added to readings and measurements. The only limit to measurement was the quality of instrumentation, which was constantly improving.

Time was thought to be continuous, absolute, infinite, and entirely distinct from anything else. It elapsed at the same rate throughout the universe and was not affected by gravity or the motion of physical objects. Like space, it was infinitely divisible: there was no absolute unit of time. Like space, it was something that everything was *in*: a container unaffected by its contents.

Matter occupied space and passed through time. In the early 1800s it was considered by many to be of a continuous, infinitely divisible substance, but by 1900 it was almost universally accepted as consisting of discrete particles, or atoms. Matter existed without any thought of dependence on conscious experience. The amount of matter in a physical object was measured by its *mass*. Mass could be detected by the force an object produced in a gravitational field (its *weight*), but also by its mysterious tendency to resist acceleration. The more mass an object had (even outside of a gravitational field) the harder you had to push to get it going. Nobody knew how matter was able to do this—how it extended its claws into space to keep from changing its motion—but there were some ideas. This resistance of mass to acceleration supported the concurrent assumption that space was fixed and absolute. Space was something "out there" in relation to which massive objects resisted changes in motion.

This was the world Max Planck grew up in. He was extremely impressed by the grandeur of natural law, and wanted nothing more than to do his part to confirm it. He was uninterested in new ideas or explanations. As a student he told a university professor that he was considering physics as a profession. The professor told him that, "In this field, almost everything is already discovered, and all that remains is to fill a few holes." Planck responded that he did not wish to discover new things, only to understand the known fundamentals of the field.

In the year 1899 Planck was a forty-two-year-old theoretical physicist at the University of Berlin. He had become interested in a phenomenon known as *black body radiation*, or the energy given off by an object that absorbed all of the radiation it encountered (*black* meaning that it does not reflect any light). In a description of the radiation a black body emits he found that its energy was not continuous, but came in a form restricted to whole number multiples of an elementary unit, that is, in discrete pieces. He established a relation between energy and wavelength for such a unit in the equation:

$$E = hn$$

where E is energy, n wavelength, and h a constant.

This is the famous Planck constant, equal to 6.626068×10^{-34} m^2 kg/sec: an extremely small number. But the fact that it is not zero has revolutionized science. If it were zero, energy would be

a perfectly smooth, continuous plenum, as it was always assumed to be in classical physics. What he had discovered was
that energy comes in tiny bundles, or packets. Planck had discovered the *quantum*.

Planck dismissed the quantum as "a purely formal assumption ... actually I did not think much about it..." He did not like
what he found and searched for alternative explanations. He
later described his discovery of the quantum as "an act of de-
spair ... I was ready to sacrifice any of my previous convictions
about physics... My unavailing attempts to somehow reinte-
grate the action quantum into classical theory extended over
several years and caused me much trouble." A full description of
what the quantum was and how it worked did not come about
until the 1920s. Even now, we have but a sketchy understanding of what the quantum really is.

Five years after Planck's discovery a younger contemporary, Al-
bert Einstein, found that light, too, is quantized. Quanta that
we *see* are called *photons*. $E = hn$ applies to a photon, where *n* is
its wavelength, or *color*. Light, Einstein discovered, is not a continuous plenum, but comes in tiny dots of color.

Today, we still do not thoroughly understand the quantum but
we know it is real. Thousands of experiments over the last hundred years have shown time and again that the universe and all
the objects moving through it look sharp-edged in everyday life,
but they are actually grainy and rough-edged, if you look very
closely. Physical reality is somewhat like a newspaper photograph: from a foot or two away images look distinct and
smooth, but if you look up close you see the photo consists of
thousands of tiny ink dots. There is no clear point at which an

image ends. If you keep your nose up to the page all you see is a mish-mash of ink blobs. The picture re-emerges from the dots only when you move away from the page and re-connect the dots. The real universe is like this. The Planck constant is extremely small and the "dots" are much smaller than anything you can see with the naked eye or even with an ordinary microscope, but they are there. The picture you see of physical reality is a connection between the dots.

Quanta are hard to understand because they are fundamental units not of matter, but of energy. (Actually, they are units of *action*, or energy times time.) Energy is not a substance or an object you can see or touch. By itself it has no length or width. It is something you see evidence of in the behavior of physical objects, in how they move, radiate, change, disintegrate, and bond together. Energy quanta show up in the behavior of subatomic particles. Electrons, for example, can occupy only certain distinct "quantized" energy levels in their "orbits" around atomic nuclei. They exist only at distinct distances from the nucleus: at level 1 or level 2 or level 3, but never anywhere in between. There is no such thing as level 2 ½: there is no space there. They may take a "quantum leap" to a higher or lower energy level, but they never occupy the space between one level and another. They don't even pass through it. It is difficult, therefore, to picture what a quantum of energy would look like. You might think of it as a little particle, but a particle of what? It is more like a little piece of space, or time, or mass, or all three squeezed together. It is more like a little piece of perceptual consciousness.

After many years of trying to understand quantum physics, another physicist, Werner Heisenberg, discovered that it is impossible to know a particle's momentum and its location at the same time. The mathematical relation that bears his name is:

$$\Delta p \times \Delta x \geq h/2$$

where p is momentum, x space, and h the Planck constant. (The Δ stands for change, variation, or uncertainty.) This formula states that uncertainty in momentum multiplied by uncertainty in location is always equal to or larger than $3.31 \times 10^{-34}\,\text{m}^2\,\text{kg/sec}$. This is an uncertainty built into the universe. It is always larger than something extremely small, but that it exists at all means that there is a distinct limit to the dimensional world. No matter how small h may be, the Heisenberg Uncertainty Principle states that space-time is not absolute. Space-time is broken up into little quanta. Objects cannot be located within individual quanta; they are located only in the relation of quanta to each other.

Momentum, the p in Heisenberg's relation, is equal to mass times velocity. The relation says that the more you know about a particle's mass and velocity, the less you can know about where it is, and vice-versa. If you know exactly where the particle is you can know nothing about its mass and velocity. If you know exactly its mass and velocity, you can know nothing about where it is. This is not a matter of measurement; it is a limitation of the fabric of space-time and, interestingly, of *mass*. Mass, it seems, is also a structural component of the quantum. On the macroscopic level space, time, and mass are distinct and separable, while on the quantum level they melt into one another. On the everyday level of tables and chairs, space, time, and mass are separate components of things like momentum and velocity—we can say which is which—but on the more fundamental level we cannot. Space, time, and mass are not fundamentally different.

On the everyday level, velocity equals space divided by time (miles per hour or meters per second), and momentum (p) equals mass times velocity:

$$p = g \times m/sec$$

where *g* is grams, *m* meters, and *sec* seconds. We can determine an object's momentum if we measure these three dimensional components. A baseball weighing 10 kilograms traveling 100 meters in 2 seconds has a momentum of 10 × 100 / 2 = 500kg-m/sec. As long as we can measure each component accurately we can put them together and derive the baseball's momentum. But on the quantum level momentum is not a derived quantity. It is not built of separable parts. The momentum of a subatomic particle cannot be broken down into separate dimensional components. Again, space, time, and mass blend into one another below the quantum level and cannot be pulled apart. To say the same thing another way, space, time, and mass come into separate existence only above the quantum level.

According to the Heisenberg Uncertainty Principle, quanta are too coarsely interconnected to know much about the size or motion of anything that fits between them. And they are not evenly spaced: the closer they are in space or mass, the farther apart in time. If quanta are like tiny points in a universal matrix of space, time, and mass, then what a small object looks like suspended in such a matrix depends on the angle from which you view it. The questions you ask determine what you see.

The Heisenberg Uncertainty Principle is sometimes explained in terms of the *observer effect*. The act of observation impacts that which is observed, which is to say that the objective world is not independent of conscious experience. But the explanation itself is objective: to see a particle you have to bounce a photon off it, which, because the photon has momentum, disturbs the particle in some way so that you no longer know where it is. If you use a short wavelength (high energy) photon you can say a lot about where the particle is because the short wavelength limits uncertainty in space, but you cannot say much about the particle's momentum because you have hit it so hard with the photon. If you use a longer wavelength (lower energy) photon, you don't hit it so hard and can say more about its momentum, but less about its location because of the longer wavelength. There are limits to the usefulness of this explanation, for a variety of reasons, but it is one way of trying to "visualize," from the outside, what is happening.

Louis de Broglie found that particles consist of *matter waves* that are spread out in space. The slower and smaller they are (the less their momentum) the more spread out they are. This means that subatomic particles cannot be said to exist at exact locations. De Broglie knew that light, though massless, behaves like a particle at times and reasoned that a particle should also behave like a wave at times, even if it has mass. His idea was speculative but later verified experimentally. It showed that quanta are grainy not only in space and time but also in mass. Extremely small objects do not have distinct edges in space, time, or mass. Even larger objects like tables and chairs have blurry edges, though the effect is so much smaller that it is virtually immeasurable.

De Broglie showed that as subatomic particles coalesce into larger objects their wavelike nature diminishes. As particles clump together, their collective momentum increases and their de Broglie wavelength decreases: the uncertainty of their location decreases in space, time, *and in mass*.

Erwin Schrodinger said that the wave describes the *probability* of a particle being found at a certain point in space. You can never say for sure if the particle will be at some particular point, but the wave tells you exactly the chances that it will be there.

Uncertainties, waves, and probabilities are all ways of trying to get at what the quantum world is all about. But what is really happening? What is meant by a *wave*? Is it the way a particle moves? The way it bounces off other particles? What does a wave have to do with probability?

Imagine your house with sensors embedded in every square meter of the floor. As you walk about from room to room they keep track of where you are by registering a reading once every minute of the day. After a period of time you assemble the data into a graph that shows each room of your house and a dot for each minute of the day. Certain places, like your favorite living room chair, your bed, and the square meter in front of the kitchen sink, will show high densities of dots, while other places, like closets, stairways, attics, and hallways, will show only a few. You can see something about what your day is like by the spread and density of the dots throughout your house. Depending on how you set up the graph, you can make the dots form a wave. If you arrange the rooms in your house into line segments, set the segments one after another on the x axis, and then represent the number of dots in each segment along the y axis, a wavy line will result that shows how often you are in various locations. You may wish instead to show the whole floor plan of your house and let the density of dots in each square meter reveal the wave pattern. Or you may find some ingenious way to show the basement, the second floor, and attic as additional dimensions, or you may even find a way to include time, and show yourself moving around the house. The shape of the wave depends on the parameters you use to represent it. The picture you see of yourself over a day's time depends on how you set up the graph.

This is the "wave function" of your daily life at home. It may include repeating patterns at certain times of the day or days of the week that form an oscillating wave, like a sound or water wave. It may show you in the kitchen at the same times every day, or taking out the trash on Thursday nights. But the wave is not material and you do not see it at any point during the day. There is no way to see or touch it in the house. It is "there" only as a relation among dots. It is no more than an invisible pattern that develops from how you move, that shows something about you. You could turn the wave into a probability by counting the dots in any particular square meter and dividing by the total number of dots. If, for instance, there are 13,000 dots in front of the television set every 100,000 minutes, you can say there is a 13 percent chance of finding you there.

Waves and probabilities are used to describe the behavior of quantum particles like photons or electrons because there is no other way to describe their behavior. In your case, waves and probabilities describe *you* as you move about the house, but in the case of quanta, we do not know what they describe. We don't know *what* is moving—all we have are the waves and probabilities. It is as if all we knew about you were the dots on the graph, as if there were a dot at the front door at 11:06, and two more down the hall at 11:07 and 11:08, but no knowledge of *you*. It would look like you were walking down the hall, but that's just something we've added on top of what we know for sure. We're not certain just where you were between the dots, or *if* you were. Because the sensors are spread out in your floor, and because we are taking readings only every minute, our knowledge of what goes on in your house is *quantized*. You could be running all kinds of patterns between 11:06 and 11:07 that don't show up. This is what real space and time are like for electrons and photons.

If we wanted better information about your movements, we would put sensors in the floor every square centimeter instead of every square meter, and take readings every second instead of every minute. We would improve the space-time grid in your house by bringing its Planck constant down a few notches. Now you will show up not as a single dot here or there, but as a *range* of contiguous dots. We will get a much clearer picture of you and how you move because the dots will be closer together and there will be more of them. But the shape you make with the dots will still move across the graph in increments—individual sensors will turn on in front of you and turn off behind you as you move down the hall. The range of dots will remain imperfect and unsmooth, and your house will still be quantized. If we install more sensors and take readings more often, the range will get smoother, but because we cannot add an infinite number of sensors we will never have a perfectly smooth image. As the range of points becomes smoother and smoother we become more interested in its shape and less interested in the wave. The range of points that creates the shape is so much larger than individual points that the wave disappears for all intents and purposes. This is what happens in real space and time for objects larger than electrons and photons.

Classical physics assumed that there were an infinite number of sensors or, what is the same thing, that the world we see was not a picture at all. It did not know that physical objects are really ranges of points. Images are so smooth at the level of tables and chairs that physicists did not notice the composite, quantized structure of perceptual experience. The quantum is so small it is a wonder Planck ever discovered it.

The discovery of the quantum is like running the scenario in your house in reverse order. At first, there seems to be an infinite number of sensors embedded in the floors and the range of points you make as you move about seems perfectly smooth and contiguous. It is so smooth on the graph you use to interpret your data that no one notices the points at all. The reality of *you* is the shape taken by the range of points. But as we reduce the number of sensors and the frequency of readings, the picture you make on the graph begins to look discontinuous and grainy. Spaces begin to show between points. This is like looking at smaller and smaller particles in the fabric of space-time. There is still a shape to the range of points, but the shape moves incrementally by points turning "on" in front of it and "off" behind it. Finally, as the sensors are back to one per square meter of the floor space, you disappear between the points. There is no range of points between points. Now all we know about you is a probability of where the next point will be. All we have now is your wave function.

We know from the wave that you are more likely to be at the sink just now than in bed or in front of the television, but we do not know for sure.

We cannot add more sensors to real space-time and we will never have more accurate locations for photons or electrons because there aren't any. We will have to depend on their wave functions. But there is something we do have for quanta that we do not have for you. You move all over the place, depending on what needs doing around the house. You water the plants, answer the phone, stir the soup, and empty the clothes dryer as the need arises, so your wave pattern is not the same from hour to hour or day to day. There are repeated patterns, to be sure, but no one day is entirely like another. Quantum particles, on the other hand, are more dependable. They have stable wave patterns. Each electron, for example, is absolutely identical to every other electron and has only one wave pattern for a given physical state. Its wave function does not say where it is or what it will do, but it does say exactly what a *group* of electrons will do. If allowed to pass through a tiny hole, even *one at a time*, electrons will land in a pattern on a photographic plate exactly as described by the wave. They are not bouncing off each other—each is its own wave. We don't know what each will do, but we know what they will *all* do: 13 percent of them will fall at a specific point the way you will be in front of the television 13 percent of the time. The wave function, therefore, describes both how a large group of electrons will distribute themselves and the probability of finding any one electron at a particular point. We know what the uncertainties are and how they interact with other uncertainties and there is a *statistical* basis for cause-and-effect relations. We cannot speak of the mechanics of any one particle but we can speak of *quantum wave mechanics.*

Wave mechanics is a lot like marketing analysis. In an economy of a hundred million people marketers can predict with some certainty, say, that a million consumers on any given day will buy a bag of potato chips. Inversely, they can tell that the probability of any one consumer doing so will be one in a hundred. But they have no idea what any *particular* consumer will do and don't much care. The reality that marketers respond to is not what Mrs. Smith or Mr. Jones might do tomorrow, but to the invisible potato chip wave function that hovers over the economy as a whole.

To extend the analogy, marketing wave mechanics would be the study of how the popcorn wave function interferes with the potato chip wave function. We still don't know what Mrs. Smith will do, but we can say she is 30 percent less likely to buy a bag of potato chips if there is popcorn sitting on the shelf next to it.

In quantum mechanics, the wave function describes the uncertainty of a particle's location between observations. It describes, for instance, the uncertainty in where an electron may be after it passes through an aperture but before it lands on a photographic plate. Once the electron lands on the plate the wave function "collapses" because we know exactly where it is. There is no longer any uncertainty once an event happens and becomes the past. The wave function collapses only as the electron is observed and becomes a quantum point in the matrix of space, time, and mass.

Physical reality consists of tiny bits of energy. Bits come in the form of quantum particles such as photons or electrons at discrete energy levels. What we see, hear, and feel around us are shapes that these bits take to become objects of perceptual consciousness. They are so small that the shapes look smooth and continuous on the everyday level.

Below the quantum level space, time, and mass do not exist separately. The uncertainty principle describes the impossibility of knowing a particle's momentum and location at the same time: this is another way of describing the dissolution of dimensional structure at the quantum level.

The dimensional structure of what we call physical reality arises only above the quantum level. Because the parameters we use to understand physical reality do not exist in distinct form below the quantum level there is no way to understand what is going on in any normal manner. Physical objects—tables and chairs and other things that are in space and time and "have" mass—are shapes created by aggregates of quantum particles above the quantum level.

We are used to thinking of such objects as existing "out there," in the dimensions, whether we are looking at them or not. But quantum particles exist only when observed, and there is no way to show that aggregates of quanta exist other than when observed. What we call physical reality is nothing other than perceptual consciousness. The space, time, and mass structures of physical reality are nothing other than the dimensional structure of consciousness.

The shape a wave takes—or the shape taken by a range of dots—depends on the parameters of the graph used to interpret the information. What *you* look like as you assemble data from the sensors in the floor of your house depends on what kind of graph you use. You could use a pie chart or bar graph to show percentages of time spent here or there, or you could plot points along axes on graph paper, or even in a three-dimensional holographic representation.

In the physical world the parameters are space, time, and mass. That is the graph paper of physical reality. But paper shows only two dimensions. To include time, for our purposes in this book, we will choose a new model for space-time. We will switch from paper to a screen with little lights that turn off and on. To add more space, we will make the screen three-dimensional. We will call it the *photon screen*, and it will become our model of the four-dimensional physical reality that you see. When we add mass, it will become the five-dimensional *quantum screen*, our model of perceptual consciousness as a whole.

The quantum is the fundamental unit of perceptual experience. It is everything that you see, hear, smell, touch, and taste. Yet it is too small to notice in everyday living. It shows up in a big way only in physics laboratories, with the help of cloud chambers and particle accelerators. Or in a quiet mind.

QUANTUM
MEDITATION

Quantum meditation is with eyes open.

Opening the eyes exposes the mind to a barrage of distraction and confusion. Most people who meditate will tell you this. It diminishes concentration and reduces the chance of achieving liberation from earthly attachment. It is not recommended. Pull the arrow out and continue on your way. If you would attain liberation in this lifetime, do not look at the arrow.

Quantum meditation is looking into space the way Buddhist meditation is looking into the body. It is the same cultivation of awareness, the same balance of concentration between wholeness and detail, and the same detachment from self. It should not be practiced exclusively. It is best to learn concentration with the eyes closed and open them later.

Quantum meditation, despite its New Age ring, is not a cult or school of practice; it is, in fact, no more than the words I give it here. It is an application of the Buddha's method to an area of life that he did not know. He could not have intended for meditation to be used to explore the quantum, and there is a danger of damaging the tool. But we do not apologize for using it this way. We will be respectful of the sacred tradition that has grown up around the Buddha's method, but we do not apologize. We will use the method carefully, for a new purpose.

We will use meditation to analyze consciousness. We will see that multi-cellular consciousness is a composite of cellular consciousness, separable into perceptual and observational realms, and dimensionally structured on the macroscopic level. We proceed from beyond the bounds of science to see what science is.

Since the Buddha developed his method we have come to understand that all living things are made of cells. Trees, horses, insects, humans, bacteria, and earthworms consist of microscopic units of life, each with its own digestion, metabolism, defense, and reproduction. Types of cells and the order in which they are arranged constitute the larger organism. In a reductive sense, a tree or horse is *no more than* the collective of its cells— the organism is reducible to the sum of its parts. But we know that life does not work this way. Life is a wholeness that dies when parts are separated. A multi-cellular organism is a community of cells with an existence over and above that of each cell individually. It is a being of its own. As multi-cellular organisms ourselves, we are the wholeness of the cell community that floats above the separate cells of which it consists.

In Buddhist meditation, looking at the body from the inside, you see the world of cellular being—pressure, stress, pain, taste, soothing, relaxation, digestion, the solidity of bones and the stretch and ache of muscles. You become aware of the sensations cells experience. You do not experience what they experience but you are in their world. It is a world without seeing, hearing, or smelling—a world without the higher sense organs.

There is no information here that cannot be read by a single cell. There are no objects beyond the body. This is the world of the chemical and tactile realms of perception—a world without space. It is the world experienced by single-celled organisms, of plants, and of cells in the bodies of animals and people. Sound and light do not exist here, but this is the elemental world from which sound and light are built, and therefore, the universe itself. Organic consciousness is built from cellular consciousness, and you must know the parts to know the whole. You must know the world of the cell to know the larger world. This is the world with eyes closed, the world away from sounds and smells, the world of the body.

There is a rudimentary sense of space in this chemo-tactile world, but not the dimensional space of multi-cellular consciousness. The body has an up and a down, a right and a left, but images in the body are not distinct. There is no clear point at which they begin and end and nothing fully separating one from another. There are no sharp angles, shapes, or appendages. Images at one place in the body affect the body as a whole, occupying it entirely to some extent. There is no clear separation of subject and object. Gross sensations—arms, legs, fingers, and muscles— remain identifiable despite the lack of a clear spatial context.

Gross sensations come to the brain as impulses from aggregates of cells. The brain is not sensitive enough, in ordinary circumstances, to hear the voice of an individual cell. It takes many thousands of them, perhaps millions, to be noticed. Nerves spreading throughout regions of the body pool impulses and package them in a manner that seeks the attention of the mind: the more acute the need for attention, the higher the intensity of sensation. The body lets you know in this way that you should attend to its needs. When you react to a gross sensation by moving a body part, you allow the wholeness to become the part. Awareness is diminished to that extent.

If you do not react, but continue listening to what your body is experiencing, you allow the parts to become the whole. Your cells become one: at the moment you see them individually you see them as a whole. Maintaining wholeness against the pull of sensation leads to transcendence of the parts. You see through gross sensations to the subtle sensations they are made of, through aggregates of cellular sensation to cells themselves. You experience kalapas. Kalapas are the sensations of individual cells. They are not you. You are no longer you when you feel them. This is a state of consciousness difficult to attain even by those who meditate regularly.

A kalapa is the voice of a cell.

When you open your eyes, subtle sensations in your retina become points of color in space. Kalapas become light quanta. The cells of the retina become the photon screen.

A photon is the voice of a cell, in space.

Space arises from nowhere.

Quanta are messages from individual cells built into a space-time grid. The grid is the context of multi-cellular experience, the template on which quanta are arranged. Quanta are discontinuous on the fundamental level, as living organisms are discontinuous on the cellular level. Intervals between one quantum and another correspond to intervals between one cell and another. The discontinuous energy bundles Planck noticed in 1900 were kalapas nestled in space.

Kalapas are the limit of subjective experience: when you feel them you are no longer there. Quanta are the limit of objective experience: where you see them is the end of physical reality.

How do subtle sensations become quanta? How does what you feel in your body become the larger physical world? The step intermediate between a kalapa and a quantum is a photon. It is opening your eyes and seeing light.

Light is visual consciousness.

When you sit in meditation with eyes closed you notice that the intensity of sensation varies from one part of the body to another. You feel weight in the pelvis, tension in the knees and feet, and restlessness in your hands, while you may feel nothing in the abdomen or in the space between joints. You may also notice that there is a lot of "room" in front of your eyes. You "see" lots of subtle sensations here popping in and out of existence as they vibrate across what would be your field of vision were your eyes open. When you do open your eyes, it is these subtle sensations that become photons.

With your eyes closed and your mind still you feel the body. Aggregates of subtle sensations clump together to form shapes of gross sensation—arms, legs, aches, pains, breathing, etc.—against a background of disassociated subtle sensations. When you open your eyes, a million light quanta fill your field of vision. This is the photon screen: the visible realm of consciousness. Each pixel—each photon—is a point of color in three dimensions of space and one of time. Visual objects are aggregates of photons— gross sensations—arranged in contiguous dimensional patterns. The tables and chairs you see around you are ranges of color shifting across the space-time of the photon screen.

Cells in the retina do not see. Like other cells, they are capable only of chemical and tactile perception, and like other cells, they relay experience to the brain through the central nervous system. Unlike most other cells, however, each retinal cell is in contact with its own neuron leading directly to an optic lobe of the brain. The optic nerve is a bundle of a million neurons, each connecting an individual rod or cone cell to the brain. Each sends an electrical impulse describing what it "feels." The brain arranges it with impulses from hundreds of thousands of other retinal cells into patterns that become visual consciousness. Cells *feel*; you *see*. This is the dual nature of light.

The brain cannot see anything smaller than a photon. It cannot tell the difference between seeing and feeling at the level of individual photons. There is, therefore, no distinct visual realm of consciousness below the quantum level, and no corresponding space dimension.

If you look too closely—if you look at an object that is too small—space dissolves back into cellular experience. Quanta go back to kalapas. Even with the most powerful microscope you can see nothing smaller than a quantum because the context of visual experience is too grainy. Cells in your retina use combinations of quanta to tell the brain what they feel; they cannot, therefore, tell the brain about anything smaller than a single quantum. Within a single dot there is no connection between dots.

If the quantum were as yet undiscovered we could surmise its existence from the cellular basis of biological function. When an organism experiences anything, some or all of its cells experience something at the same time. *What* the organism experiences may differ categorically from what its cells experience, but multi-cellular consciousness has to be a composite of unicellular consciousness. It has to be a structural relation of sensory input from individual cells. It must, therefore, be quantized at some level. It has to be reducible to constituent parts. Our experience is our cells' experience, though we draw a larger picture from it.

Space is a construct, a means by which information is assembled into a larger mental picture. It is a structural context, a pixel screen on which cellular input is arranged into perceptual images. Space is like mortar. It makes a pile of bricks look like a building. Where is the building in one brick?

The Space Dimensions

Space is created from time: One second of time equals "c" meters of space. Space coordinates seeing, hearing, and smelling with sensation in the body. It is a way of experiencing information from three additional realms of sensation at the same time. Objects in these realms are usually at a distance from the body—not experienced in the chemical or tactile realms—but their position in space shows how they are potentially experienced in the body. Where and when objects are seen or heard indicates what must be done with the body to touch or taste them.

A sensory potential is an infinite expanse of possibilities within which actual sensory information is located. The inter-coordination of sensory potentials into space-time is perceptual consciousness as a whole.

Quantum patterns—gross sensations in space—are experienced only in multi-cellular consciousness. If you are able to experience quanta as kalapas, space will disappear.

Science did not know what to do with the quantum because the scientific tradition before the twentieth century assumed that space was a given—something already there, in the world, requiring no explanation. It was a fundamental structure of an external universe, something that underlay everything else—something that anything real was in. There was not much use talking about it, and you had to assume it was there to do science. But scientists discovered that space at extremely small dimensions does not do what it does on the macroscopic level. It falls apart. There are gaps between one point and another, gaps that open and close depending on time and mass. The clearer the space becomes, the foggier the time and mass. Worse for classical physics, what happens on the quantum level depends on someone watching it happen. Conscious experience entered into what was always thought to be purely objective phenomena. This was not something science wanted to see. Most scientists were uninterested in metaphysical questions and preferred to keep assuming that space was fundamental and absolute. But they could not. Evidence was piling up that reality goes deeper than space.

The visual world and the chemo-tactile world from which it arises are two separate worlds. How are they connected to one another, and to the auditory and olfactory worlds? Vision is only one realm of perceptual consciousness; how does one realm become the structural background for the two chemo-tactile realms (tasting and touching) and also for hearing and smelling? Why do the five realms have to be coordinated into a single world?

Vision is unique among perceptual realms. Space is a special structure built into visual consciousness from time. One second equals c meters. But space-time is the structure of *all* perceptual consciousness, not just vision. How does this come about?

The mind might have chosen, at some point in the deep evolutionary past, to use a realm other than vision to bring the five perceptual realms together. It might have chosen a dimensional structure based on hearing or touch. It did not, we presume, because light is more refined than other forms of perceptual information. It has so wide a potential—so much space—that it can be far more intricately detailed than other realms. Visual images appear at distinct locations in relation to the body and in relation to each other, making them easy to separate, label, sort, and categorize. Sound and body images are not so discrete.

Without the clarity that visual space provides, one image would appear on top of another, or blended into something next to it, and we would be unsure of what we were looking at. This would likely happen if, for the purpose of setting a common denominator, information from the visual realm were reduced to sound or tactile perception. Light, with its greater capacity to arrange information, is a better common denominator for all five perceptual realms. The richness and clarity of organic consciousness is the richness and clarity of light.

What you hear and smell is not light, but the olfactory and auditory information reaching you is structured so as to be *coordinated* with light. The non-visual realms are experienced in a dimensional context that is the same as that of vision. Objects you hear or smell you may or may not see, but *where* you hear or smell them is where you will see them if you look. The structure of light is the structure of space-time itself.

Space is the way quanta are arranged. The three spatial realms—seeing, hearing, and smelling—are the three specific to multi-cellular animals. Each corresponds to space dimensions that serve as potentials for actual sensory information.

Information, to be information, has to be an actual within a potential; it has to be something that could be something else: a dot that could be a dash, a yes that could be a no, an on that could be an off, or a number that could be some other number. That is what makes information a form of consciousness. A number or a signal without a potential is not information. The number *eight*, for instance, tells you nothing if it has no context. But if you see it at a particular location on an overhead scoreboard in the high school gymnasium, it tells you something that will make you either happy or disappointed. The part of the scoreboard within which you see the number creates the potential that turns the number into information. The English alphabet is a twenty-six-letter potential that turns lines of ink into information. Each letter could be some other letter, and that is how it tells you something. Similarly, a dimension is a potential for perceptual information. Each realm of perception is associated with one or more dimensions. Space-time as a whole is the potential for perceptual consciousness as a whole. The dimensions are coordinated in such a way that where and when you see an object is where and when you potentially touch it, whether you actually touch it or not.

Actual perception within dimensional potentials is the vibrancy of perceptual consciousness. It is what makes you feel alive and wondering what will happen next. It is the potential that gives information meaning.

Consciousness is not what you see; it is the potential to see.

Other realms of perception are coordinated with vision by an extension of the photon screen in all directions beyond the visible. This is the quantum screen.

The photon screen and the quantum screen are models I use to explain perceptual consciousness. They are like video screens in many ways but have no objective existence in space or time. But neither are they imaginary: you see the photon screen whenever you see anything, and hear, smell, taste, or feel the quantum screen whenever you hear, smell, taste, or feel anything. They are always there.

The quantum screen coordinates everything you perceive into a grid pattern that helps you make sense of the world. It is, in fact, the structure of the world. Information from all realms of perception is coordinated into it in such a way that you hear or touch objects at the same location in space and time that you see them. You know exactly where and when you will touch tables and chairs because you see them there. It is this coordination in space-time that gives physical objects an illusory sense of independent existence, and where the concept of material substance originates.

If you are not looking at an object where you hear or touch it, you will see it there if you do look. Where you hear or touch it there is *potential* vision. If you are not touching a chair when you see it on the other side of the room, you will feel it there when you walk over and reach out your hand to that location in space. The quantum screen is mostly potential information; where you actually perceive an object in any realm is where you potentially perceive it in all realms.

Even the body finds its way onto the screen. In meditation the body can be experienced without the quantum screen (without space), but when you open your eyes, tactile perception is coordinated with light—with vision. You feel your hand where you see it—in space. The tactile realm becomes a second time dimension foreshortened on the screen—the per second *per second* of acceleration. (You do not feel uniform velocity—the first *per second* of your motion—you feel only *changes* in velocity: bumps, curves, speeding up or slowing down.) The kinesthetic sensations you feel as you move your body about are motions in relation not to space, but to space-time. When you are sitting in an airplane that is taking off down the runway you feel a uniform sensation throughout your entire body that is exactly proportional to the acceleration. This is a simultaneous activation of kalapas—of all quanta on the screen.

The second time dimension—the dimension you feel—is the dimension of mass. The mass of a physical object is its resistance to acceleration, as measured in the second time dimension. All physical objects are extended in mass, time, and three dimensions of space. The quantum screen is, therefore, five-dimensional where the photon screen is four-dimensional. You feel mass on the quantum screen as tactile sensation; you see it there only indirectly, as resistance to acceleration.

The point where kalapas emerge from the background of unstructured consciousness is the point where separate realms of consciousness begin to take form. Sensations are distinguishable from thought in that they last for a while. They are extended in time. Thoughts can be associated with points in time, or series of points in time, but body sensations *last* through time, even if briefly. The ache you feel in your left ankle is still there; you can go back to it. It may grow, shift, change in some way, and eventually disappear, but it exists through time. You can watch it arise, endure, and pass. Thoughts, on the other hand, pop in and out. You might be able to find them again but you have to reach back in. The difference is subtle, but important. In fact, body sensations are not only in time, they *are* time. The time dimension arises as a means of distinguishing the body from thought. The realm of tactile sensation arises as a separate category of consciousness.

The realm of chemical perception arises in a similar manner. Chemical sensation is more fundamental than tactile in that it involves processes within cell membranes, not only on their surface. Cells depend on a constant intake of chemicals through the membrane and live by chemical reactions that take place within. There is enormous survival value in developing tactile perception as a means of screening molecules on the outside before allowing them in. But cells have to know the difference between "feeling" and "tasting" and need to develop separate but interrelated categories for each. Cells in your body may not experience these realms in the way you experience them, but they have ways of making you aware of what you need to know. The sense of taste is a specialized function of the chemical realm that monitors ingestion, but it is not the only place you experience chemistry. Chemical sensations are experienced throughout all parts of the body, particularly along the digestive tract. Much of your emotional experience is based in chemical perception that is not specific to your taste buds.

The chemical realm becomes distinguishable from conceptual, or non-dimensional, consciousness through a time dimension. It must be distinguishable also from the tactile realm, and therefore falls into a separate time dimension. This is all but impossible to experience directly, even in meditation, and becomes apparent only when the relation between perceptual realms and the overall structure of consciousness is fully appreciated. It is surprising, on the surface, that so fundamental a structure of the universe as time would be identified with so insignificant a human sense as that of taste, but we are dealing here with the rise of a category of consciousness that goes back to the formation of the first cell membranes, and earlier. Taste buds may not be a large part of the world, but biochemistry is.

In terms of biophysics, the difference between the chemical and tactile realms is the difference between bound and free electrons. Bound electrons—those associated with atomic nuclei—

are the basis of all chemical interactions, biological and otherwise. How they interact, and how they are shared by atoms within molecules determines the nature of all chemical substances and how they react to each other. Chemical perception within cells is as rich and diverse as chemistry itself. Free electrons—those that escape atomic nuclei to form static charges and electrical currents—are what you feel in the tactile realm. What you feel when you touch something is an impulse of electrons passing through nerve cells to your brain. (The impulse is triggered by repulsion between outer electrons in the object and those in your body.) The "matter wave," or frequency, associated with the quantum nature of free electrons is a time dimension separate from that associated with chemical sensation. This frequency is in the second time dimension. It is also the second *per second* of acceleration that your body feels.

Subtle sensations are free electrons. You feel them as points in time. But they are not extended in time, as are the shapes they form. We are more interested in them than in chemical sensations because it is on top of them that the auditory and visual realms are built.

If the space that comes with light arises in the brain, where is the brain? Where are the cells that experience the photons that create space?

There is a circularity of reasoning in what I have said to this point. Space seems to come into being somewhere in the optic lobe, as kalapas become quanta. But where were the kalapas before? Where were the retinal cells and the optic lobe? Where was the light before it came into the eye? I have assumed the existence of space in attempting to explain it.

The problem is the picture. If you picture a photon (a light quantum) crossing the room, entering the eye, and striking the retina, the conceptual tools you are using require space. It is difficult or impossible to picture it any other way. Dimensional concepts are what you have available to make sense of mechanical (cause and effect) relations, but they do not work below the quantum level because space dimensions themselves become distinguishable only above the quantum level. You cannot picture what is going on below because there is no space there to have a picture in. You cannot properly picture a photon flying through space because it is only with the experience of the photon that space comes into being. All that is actually there is a point of light in relation to other points of light.

A photon is never in space. It is a point of color that creates space in relation to other points of color.

·

A dimensional picture can be useful in relating one area of spatial experience to another. You can, for instance, allow yourself to picture cells in space receiving and emitting electrical impulses. But such a picture always leads to paradox when applied to quantum reality. You end up looking for the ox you are riding on in search of the ox. Scientists have been struggling for a hundred years to find common, everyday concepts to explain quantum physics, and never yet succeeded. This is less because life is paradoxical than that the attempt to understand life is paradoxical. Concepts loop back upon themselves. Meditation, in concentrating on direct experience, attempts to penetrate through concepts that overlie experience. Quantum meditation attempts to see through even the most fundamental context of experience: space-time.

The circularity centers on our understanding of single cells from a multi-cellular perspective. As a multi-cellular being, you cannot help but use spatial concepts to picture what a living cell is. You see it as a little chemical box with membrane and nucleus, absorbing impulses in one side and passing them out the other side to the brain. This is what a cell might seem like under a microscope, but it can be nothing like the experience of the cell. It is certainly not your experience of one of your own cells. The space you use to put the picture together arises only when the experience of many cells is composed and arranged on the quantum screen, an arrangement a single cell never sees. It is *you* who enwrap the cell with space as you picture it sending signals to the brain. The space is not there until you add it.

You have no way to know what actual cellular experience is, but you can glimpse its non-spatial parameters by looking into the body with a quiet mind. You can see what every cell in your body has to say.

Ultimately, to know where space comes from, you must not picture a photon sailing across the room, bumping into a retinal cell waiting in space-time. You must not picture the cell or the photon at all. To know how the photon comes into space you have to see it as it actually is: a tiny point of color among many. You have to look.

Quantum meditation is with eyes open. It is with ears and nose open, looking, hearing, and smelling objects within each realm exactly as they are. It is letting space in. You do not need it if you do not need space.

Quantum meditation is accepting the complexity of the spatial world. It is risking distractions outside of the body. It strives, with eyes open, for the wholeness and simplicity of meditation with eyes closed. With work, it finds it. The identity of quanta with kalapas is the identity of objective with subjective experience. An object at the far reaches of the universe is of the same primal substance as a point of tension in your left ankle.

Quantum meditation is a voyage through complexity to simplicity. It is a path to the unity of all experience.

But let us not step from the path just yet. Let us not yet recline in the unity of being. Instead, let us linger a while longer in complexity, wading a little farther through the verbiage in search of a way out that leaves a better sense of where we have been. We must have a better look at the arrow.

We have had a glimpse of unity. Let us look to see if it is what is there.

BODY TO LIGHT;
LIGHT TO WORLD

*E*ach realm of perception is associated with a space or time dimension. This is the simplicity of the dimensional structure of consciousness. Its complexity is the association of realms to each other.

Light evolves from a portion of the body (the retina) and becomes the structure of perception as a whole.

Part of what you feel in your body evolves into vision; the space-time of vision evolves into the space-time of all perceptual consciousness.

Kalapas become photons; photons become quanta.

Body to Light, Light to World.

When your mind is quiet, subtle sensations pop in and out of existence throughout the body. When thoughts are not pulling you in other directions you can see kalapas as bhanga, an undisturbed expanse of subtle sensation. Even if you do not meditate you can catch glimpses of subtle sensations from time to time. These are signals from each cell of your body. They clump together to form an arm or a shoulder. They may raise their intensity to become a pain in the lower back. They form shapes that endure through time and dissolve back into the subtle emptiness of the body as a whole.

You see subtle sensations in front of your eyes more clearly than anywhere else. With your eyes closed there are no gross sensations to distract you in this particular corner of the body. This is, of course, your field of vision—without the vision.

Before you open your eyes, take a quick look at what you are smelling and hearing. Smelling is not as important a function to you as it is to your mammalian relatives, but it is as important a structural category as hearing or seeing. For a human there is less in it, but it remains a complete realm of perception. If you happen to be catching wind of something just now you may notice that it is difficult to locate or identify it if you do not see it. Without moving around and sniffing the air in several places, you cannot tell what direction it is in. Hearing, though a far more important perceptual realm to humans, is only a little better in this respect. You can tell the direction of an auditory object, but without looking at it you may have trouble knowing what it is, how far away it is, or telling much else about it. If you hear a whooshing sound on the street outside the window, you may think *car*, because you have heard the sound when you have seen a car. But unless you see it you do not know what kind of car it is or who might be driving it. Objects of olfactory and auditory perception are in space, like visual objects, but until you open your eyes the clarity and detail of the spatial world is only a little better than that of the chemo-tactile world.

Now, still watching the subtle sensations in your field of vision, open your eyes. Concentrate on the whole body as you see part of it become light. This may take a few tries. The subtle sensations that you see crackling like static electricity in front of your eyes suddenly become tiny points of color arranged in space. They fall into a multidimensional grid pattern with each point of light at its own distinct location. Shapes and patterns of light form with crisp lines and sharp edges. Unlike shapes in the body, each visual pattern begins and ends at a precise location and each has its own clear identity. There is little overlap among them and there is empty space between them. Most of the space you see remains unoccupied, as if it were potential perception waiting to be filled with actual perception.

This is the photon screen. The screen is not the photons; it is the grid pattern they fall into. It is the wholeness of multi-cellular consciousness that cannot be reduced to cellular consciousness. It is not in space; it *is* space.

Light is reducible to tactile sensation—to what your retinal cells feel. Each photon is something a cell "touches." But light is, at the same time, much more than tactile sensation. It is a reality of its own. Orderly shapes and patterns of photons—what we call visual objects in space and time—take on a reality far larger than the sums of their tactile parts. They have no substance outside of experience, but become a reality over and above the experience of a few cells in your eye. The visual realm arises from the tactile, and is composed of tactile parts, but the order assumed by the parts is as real as the parts themselves. In the everyday world, it is realer.

The visual and tactile realms each have their own reality. How do you put them together into a sensible world? How is it that you experience but one world, and not two? The only physical connection between them is the tactile reduction of photons. You could keep the two realities together by continuing to see the visual realm—the photon screen—as part of the tactile realm—the body. You could try to maintain your awareness of light as "no more than" colored kalapas in the part of the body near your eyes, but you don't. You can't. No matter how hard you try—no matter how convinced you may be that light is "nothing but" tactile sensation—you cannot avoid the orderly reality of visual objects. They are right there—you see them— you don't "just" feel them. Furthermore, the shapes and patterns that you see are so clear, detailed, and distinct that there is no way you can understand them as mere gross sensations at a special corner of the body. The visual realm is so much more varied and diverse, so much wider in range and possibility, that there is no way you can sustain your experience of it as another form of tactile perception. The parts are reducible, but the whole is not. Light has so much structural potential that you will never be able to fit it within the structure of tactile sensation. Instead, you will have to do it the other way around. You will come to coordinate the tactile realm within light. This is a convolution: the tail will learn to wag the dog.

Light has so much potential—so much empty space—that it has room for the tactile and for every other perceptual realm as well. The chemical realm remains in contact with the body—it has no spatial dimension—but the auditory and olfactory realms each assume a spatial dimension within the overall structure of light.

When you hear the whooshing sound outside the wall of your house you know to walk over and look out the window if you want to see the car and who might be in it. Before you look, as you hear the car whooshing by, you may notice that the sound, though clearly in space, is nowhere within the space that you are currently looking at. It is on the other side of the wall, beyond your field of vision. In order to coordinate the auditory realm with the visual, the structure of light is extrapolated beyond vision. The empty space that you first noticed between visual objects is extended infinitely in all directions beyond what you actually see to become the potential for all perceptual consciousness. The space-time structure of light becomes the structure of the physical world beyond what you see.

Where the chemo-tactile realms are the temporal world of individual cells, the olfactory, auditory, and visual realms are the spatial world of multi-cellular animals. Information that is heard, smelled, or seen is experience of the organism as a whole and not the experience of any individual cell. Space is how animals perceive and also how they form concepts that make experience understandable. Familiar concepts of the world disappear without space. This is what happens below the quantum level—below the level of multi-cellular consciousness.

If the tactile realm is temporal, and not spatial, how does the body end up in space?

The visual, olfactory, and auditory realms use up the space dimensions. What, then, is left for the tactile realm? How is it coordinated with light?

The body corresponds not to a space dimension but to a second time dimension at "right angles" to space-time: the mass dimension. This dimension is coordinated with space dimensions, so that you feel tactile sensation exactly where and when you see or hear an object in space.

You do not need space to experience the body. You need it to coordinate the body with what you smell, hear, and see. The body takes on a shape in space as kalapas become quanta

Part of the body evolves into light and light evolves into the world. The world then absorbs the body. The convolution is complete: the tail chases the dog, and wags itself.

Without space the body is the universe.
Within space it is a perceptual realm.

This is a roundabout way of coordinating seeing and touch. It would be simpler to keep space as an external universe, with light in space. So why go to all the trouble? Why add complexity? And how do you know that space is in light, and not light in space?

How do you know?

The speed of light is constant in relation to *everything*.

The path of a beam of light defines a straight line in space.

A physical medium for light in space does not exist.

The velocity of light is an upper limit for the velocity of everything in space.

Time slows down at velocities close to light.

Space shrinks at velocities close to light.

Mass increases at velocities close to light.

Light is either a particle of something or a wave of nothing, but not both at the same time.

That is how you know.

Dr. Einstein will explain.

EINSTEIN

The speed of light is fast but not all that fast.

If light were in space-time its physical properties would not affect space-time.

If space is in light the physical limitations of light will be those of space. The speed of light, c, will be the limiting speed for everything; c will be a limit to how many meters can fit into a second. It will be less a velocity than a structural relation between space and time. One second will be equal to c meters of space.

Space dimensions are *created* from time as the tactile experience of individual cells becomes the visual experience of a multi-cellular organism.

Einstein discovered that space is in light and not light in space, though he did not know it and would not have agreed with the statement. He never said anything about a quantum screen, though he discovered the limitations of the physical world that reveal its existence.

Like everyone else in the late 1800s, Einstein in his early years was trying to find a medium for light waves. Sound waves travel in air and water waves in water; what do light waves travel in? For a wave to go from one place to another something has to be doing the waving. There has to be a physical medium through which the wave passes. Physicists were so sure that there was a medium for light that they gave it a name: the *luminiferous ether*. Ether was everywhere. It filled everything, carrying light from across the yard or across the universe. It occupied all of space and established a worldwide reference frame for the position and motion of all objects everywhere. To learn the nature and properties of the ether would be to know the fundamental structure of the universe.

As important as it may have been, the ether remained elusive. Nobody had ever detected it. In 1887, Michelson and Morley, two American scientists, devised an experiment to fill the gap. They would use the earth itself as a moving platform. They reasoned that in its orbit around the sun the earth had to be moving relative to the ether and that its motion would be in opposite directions every six months. All they had to do was find slight differences in the speed of light as measured on earth. They built an extremely sensitive instrument that measured the speed of light in the direction of the earth's motion and also at right angles to the earth's motion. To double check they performed the experiment twice, six months apart, when the earth would be on opposite sides of its orbit. Hopes ran high that the experiment would show how light traveled through empty space and that it would establish a spatial reference point for the universe as a whole. If the ether were detected, everything would have an absolute position and velocity, and the universe could be mapped.

The experiment was a total failure. There were no differences in the speed of light between the direction of the earth's motion and at right angles to it, or between readings taken six months apart. The experiment was repeated time and again under new conditions, but no matter how many variations were attempted, or how many newer, better, and more accurate instruments were employed, the ether was never detected. The speed of light was the same no matter where the earth was or how it was moving. This was embarrassing. The ether had to be there—everybody said so.

This is where Einstein stepped in. He noted that the ether was not detected because it did not exist. There was no such thing. The concept of ether would have to go, and the worldview of classical physics go with it. A new worldview would rise in its place. The Michelson-Morley experiment turned out to be the most fruitful failure of all time. In discovering nothing, it discovered a whole new world that it was not looking for. The best science happens when you find what you are not looking for and, of course, when you have someone around to tell you what you did find.

Einstein's theory of special relativity did away with the concept of a medium for light waves. There is, as a consequence, no such thing as fixed space or absolute motion. Motion is purely relative. A train moving west over the tracks is physically equivalent to the tracks moving east under the train. The earth is a familiar frame of reference, but as far as the universe itself is concerned, there is nothing special about it or about any other frame of reference. They are all equivalent, and covariant.

But what, then, does light travel in? Light is special—it makes up its own rules. When it behaves like a stream of particles (light quanta, or photons) it doesn't need a medium. When it behaves like a wave it does whatever it does without a medium. It is not to be treated or understood like any other physical phenomenon. Here is a short list of light's unique characteristics:

- It can act like either a particle or a wave, but never both at the same time.
- Photons have no mass but do have momentum, which is defined as mass times velocity.
- Photons accelerate from zero to c in zero seconds.
- Photons always move at c meters per second. They cannot go any faster or any slower.
- There is no such thing as the passage of time for a photon.
- The speed of light sets a cosmic speed limit for everything; nothing can or will ever go faster than c.
- No matter how fast objects may be moving in relation to each other, the speed of light is the same in relation to *all of them*.

Einstein stated that the "velocity" of light is always the same relative to *everything*. This is astounding. It is true only for light. If you drive by me in your car at 100 miles per hour and shoot a bullet at 500 miles per hour in the direction of your motion, I will measure the speed of the bullet to be 600 miles per hour. This is common sense: the velocities add up. The bullet travels at 500 miles per hour relative to you and 600 miles per hour relative to me standing on the curb. But if you are zooming by me at one half the speed of light, $(1/2\ c)$ and turn on your headlights, you and I will *both* see the light traveling at the *same speed*: exactly c. I will not see it traveling at 1½ c, as you might expect. The velocities do not add up. Einstein further discovered that there is something about the structure of space and time that keeps an object from traveling faster than light. Everybody had always assumed that anything could go as fast as it wanted in empty space, but now it could cross only so many meters in a second's time, and that was it.

This changed everything. We notice the change only at a dimensional extreme, where an observer is moving relative to another observer at a rate close to c meters per second. But the change is always there, at any velocity, whether we notice it or not. Light moves at the same speed relative to everything no matter how fast things may be moving relative to each other. The true significance of Einstein's breakthrough is that this fundamental structure of light is also the fundamental structure of the physical universe.

The constancy of the speed of light means that light is not *in* anything.

Light behaves strangely in space because it is not in space.

Light is visual consciousness; space is in it.

Before Einstein, classical physics considered space and time to be separate. They comprised the framework for all physical events, but they were structurally distinct. Space was space and time was time. Einstein changed this by saying that time was a dimension, like space, and that it was inextricably interwoven with space—that space and time were essentially the same thing, related by the constant *c*. As you look across the universe or across the room, you are looking back in time: what looks like the present to you is the past for what you are looking at. The farther objects are from you, the farther into the past you are looking. An object you see now at *c* meters in the distance is one second old. The star Sirius is twelve light-years away and twelve years old. The image you see of your mother across the room is three meters distant and a hundred-billionth of a second old: you see her as she was one hundred-billionth of a second ago. Instead of three dimensions of space passing through time, Einstein brought us four dimensions of a single *space-time*.

Without getting into the specifics of Einstein's theory of special relativity we can see already that the structural relation between space and time is derived from light. We are already finding that space is a property of light.

The constancy of the speed of light changes everything because it means that without absolute space there can be no such thing as absolute *time*. More interesting still, there can be no such thing as absolute *mass*. Space and time (and mass) are components of physical reality that cannot be fully separated from one another. The only thing absolute is the relation between them. Space-time is more fundamental than either space or time.

Einstein further discovered that energy is related to mass by a factor of the speed of light squared. Why, of all things, the *speed of light*? Why times itself? Clearly, the structure of light was fundamental to the structure of the universe.

What is seen in space-time depends on how fast observers are moving in relation to one another. Einstein found that the faster they move the *shorter* the space dimension becomes in the direction of motion, the *slower* time passes, and the *greater* mass becomes. This is not just an illusion; physical reality is itself distorted by the high velocity of an observer. What an observer sees in his own *reference frame*—that is, the space, time, and mass he experiences in his car, or spaceship, or standing on the curb—does not change; what changes is what he sees in *another* reference frame. If you are in your spaceship galloping past me at one half the speed of light, *I* see everything in *your* ship shorter (in the direction of your motion), I see your watch moving more slowly, and I see that everything in your ship is more massive. To you, everything in your ship looks and weighs as it always did. As you look at me on the earth *you* see everything in *my* reference frame shorter, moving more slowly through time, and more massive. To me everything in my reference frame looks fine. We do not agree on what we are seeing, but neither of us is wrong. We are both right—there is no such thing as a privileged frame of reference. We can convert the measurements we see in each other's reference frame by what are called the *Lorentz transformations*:

$$\frac{1}{\sqrt{1-v^2/c^2}}$$

where v is our relative velocity, to come to a common understanding of what is going on. There are no objective values for any of these dimensions independent of our relative motion. It is as if you and I each carry our own dimensional reality with us wherever we go. What any observer sees depends on how fast he is going in relation to another observer. Space, time, and mass dilations appear when one observer tries to reconcile what he sees with what another sees. We never noticed the dilations before Einstein pointed them out because we do not travel around town at velocities close to c. Our reference frames are so close to one another that we cannot tell the difference.

In trying to picture special relativity it is important to remember that it is based on what Einstein called *thought experiments*. It would be very difficult to carry these out in the real world. We're a long way from building space ships that travel at half the speed of light, and a longer way still from reading tape measures, clocks, and balance scales inside space ships whizzing past at 300 billion miles per hour. (It would be hard enough to do at 3 miles per hour.) None of these difficulties matter in a thought experiment. All that matters is that thoughts are consistent with physical principles that are verified eventually. The principles of special relativity were verified eventually not with space ships but with particle accelerators. As subatomic particles are accelerated under controlled conditions close to the speed of light they become more and more massive and it requires more and more energy to get them to go any faster. They never reach the speed of light because the last meter per second would require an infinite amount of energy, even for something extremely small. As high-speed particles decay into other particles, or disappear back into energy, their time frame is slowed down at a rate proportional to how fast they are moving, exactly as Einstein said it would. Their half-lives are extended as predicted by the Lorentz transformation. The fact that Einstein's thought experiments are verified through physical experiments shows that his ideas are consistent with physical reality.

The space, time, and mass dilations of special relativity mean that the dimensional structure of the quantum screen interferes with objects on the screen at the extreme of the very fast. The screen cannot fit any more meters into a second. At a constant velocity relative to what you are looking at, you are interchanging one space for one time dimension. You are "tilting" the space-time axes of the screen. The more you tilt them, the more distortion you see in the objects. At extremely high speeds the graininess of points on the screen makes them appear closer together in some dimensions and farther apart in others. Points in space are closer together in the direction of your motion and points in time are farther apart. Points in mass become farther and farther the faster you travel, until objects moving relative to you become so massive they can go no faster. As your velocity approaches c, objects become infinitely short, their mass infinitely large, and time comes to a stop. You can never actually experience an object with a relative velocity of c because the structure of the screen reaches its limit at that point.

Objects go no faster than the speed of light because the space they are moving in is in light.

The special theory of relativity does not say that the finite properties of light make it *look like* time slows, space shortens and mass increases—it says that these happen physically. It does not say that what we see is one thing and what actually happens is something else. It says that the properties of light determine what is.

Einstein was never fully satisfied with the special theory of relativity because it dealt only with a special condition—reference frames moving at uniform, or *constant,* relative velocities. It was not a general theory because it did not deal with reference frames in non-uniform, or accelerated motion. This was not good enough.

The problem is illustrated by what became known as the twin paradox. Let us say you and I are twins. It's our thirtieth birthday. It's a nice day so you get in your space ship with a clock, a tape measure, and a balance scale, and take an afternoon spin around the solar system. As you swing around the moon and swoop down close to Earth on your way to Jupiter, you see me standing on the ground with my clock, tape measure, and balance scale. You're going at a significant portion of the speed of light, so you see my clock running slow, my tape measure shortened in the direction of our relative motion, and everything on Earth weighing more than it used to. I place a call to you and report that I see all the same dilations in your space ship. No big deal. We both know the principles of special relativity, that neither of us is right, that our reference frames are equivalent, and that when you come back to Earth we will be back in the same reference frame and everything will be normal. But then you do something impulsive. You decide to head out of the solar system, just for fun, and set a course for Alpha Centauri, the closest star. You have plenty of food aboard and a change of clothes, so you call the boss and tell him you'll be taking some time off. You open the rocket thrusters to full throttle for a week or two, coast for a few months, and relax. You get far enough into interstellar space that the sun looks like another star, and then decide you've had enough. You turn around and head back, landing on Earth a year after you left. A crowd of friends is there to meet you: it's your thirty-first birthday. But when you find me in the crowd I tell you it's my *fortieth* birthday!

How did that happen? We each saw each other's clock running slow, as expected, but our reference frames were supposed to be equal. How come I felt ten years pass when you only felt one? How come it wasn't the other way around? What was different between the two reference frames that made more time pass in one than in the other? How come you are not the one who felt the passage of ten years? Does this violate Einstein's fundamental principle that there are no privileged reference frames?

The difference is that you *accelerated* and I didn't. You and I started in the same reference frame (the earth), but you fired your rocket engines and moved *non-uniformly* to higher and higher velocities. You changed the direction of your velocity when you turned around (another form of acceleration), and slowed back down again (yet another form of acceleration) to land on the earth. We ended up back in the same reference frame, where everything looked normal to both of us, but the accelerations you experienced made your total elapsed time considerably less.

It took Einstein ten years to introduce acceleration into his general theory of relativity. Acceleration, you may remember, is identified on the quantum screen with the mass dimension. The general theory is based on Einstein's principle of equivalence, or the identity of acceleration and gravitation. Acceleration equals gravitation equals the mass dimension.

The principle of equivalence goes something like this: If you were in an elevator with no windows you could not tell if you were accelerating upward through empty space or standing sta-

tionary in a gravitational field. They are the same physically in
all respects. You would feel the same kinesthetic sensation
throughout your body in either case, and objects would fall to
the floor of the elevator in either case. You would not be able to
tell if gravity were accelerating objects down to the floor or the
elevator cable accelerating the floor up to them. Nor could you
distinguish between acceleration and gravitation by the behavior of light. If you were accelerating, a light beam traveling from
one wall of the elevator to the other would have a slight parabolic curve because the elevator would have moved up slightly
in the time it took the light to move from one wall to the other.
If you were stationary in a gravitational field, say on the surface
of a planet, gravity would bend the light beam in exactly the
same parabolic arc. (Light has no mass but it does have momentum, which is subject to gravity.)

In the general theory of relativity Einstein defined space in
terms of the behavior of light. The little parabolic arc of light
you saw in your elevator was not a curved line in straight space;
it was a straight line in curved space. Space itself is curved by
gravity, or by acceleration, in exactly the manner light travels
through space. In fact, that is the only way we know space is
curved. But if we define space in terms of light, how will we
ever find light in space?

Light, because it "travels" at the limit c, has no time dimension.
It is all meters and no seconds: a photon never experiences the
passage of time. Light, therefore, defines only space in space-
time: a line in space is whatever line a beam of light assumes. If
light is curved by the gravity associated with a massive object
like a star or a planet, space is curved exactly that way, by defini-

tion. But space-*time* is also curved, as demonstrated by objects moving at less than *c*. An object unsupported near a star or planet is *at rest* in the curved space-time of its gravitational field; it will fall into orbit around the gravitating body or fall to its surface at a constant rate in the second time dimension. Here on earth, objects accelerate to the floor around you due to our own planet's curvature of space-time. The earth is massive enough to curve space-time, but not massive enough to appreciably curve space—it does not noticeably bend light. Even the sun bends it only a little.

If we speak of the curvature of a two-dimensional surface we require a third dimension for it to be curved into. A large surface, like that of the earth, may be curved into another dimension even if we cannot see the curvature directly. If you travel in a straight line on the surface of the earth you may come back to where you started without ever realizing you were moving in three dimensions. The general theory of relativity shows that four-dimensional space-time is curved into the mass dimension near gravitating bodies. If you were to travel through curved space-time far enough in a straight line you might come back to where you started. Or, with a powerful enough telescope, you might be able to see the back of your head in the distance—or at least what the back of your head looked like several billion years ago.

The mass dimension is also part of everyday life. You feel it accelerating down the road and in the soles of your shoes standing on the surface of the earth. When you feel acceleration or gravity throughout your body you are interchanging the mass dimension for space. If it seems you are in an elevator accelerating up through empty space, all objects on the quantum screen rush by you at faster and faster velocities until they land on the floor. If you open the door of the elevator and you notice that you are not accelerating, but are stationary on the surface of a planet about the size of the earth, objects all around you will still accelerate downward until they reach the support of the planet's surface. As you step outside and feel the force of the ground on the bottom of your feet you may realize that what you are feeling—and seeing—is a permanent tilt of the quantum screen into the mass dimension. Gravity is a permanent tilt in the vicinity of a gravitating body; acceleration a temporary tilt anywhere.

Einstein showed that there was no such thing as absolute space, time, or mass. There is no fixed reference frame. He showed that a full and proper understanding of physical reality depends on what people see, that is, on observers telling us what they experience. At the same time that quantum physics was telling us to include the act of observation in understanding reality at extremely small dimensions, Einstein was telling us to include the observer at extremely large, fast, or massive dimensions.

He believed in objective reality—he never said that there is no physics without consciousness.

But that is what what he did say means.

Neither did he say that the quantum screen was based on the structure of the photon screen. But that is what he discovered.

In 1951 he wrote to his best friend, "All these fifty years of pondering have not brought me any closer to answering the question: What are light quanta?"

THE QUANTUM
SCREEN

When I was young I went to a science fair where a closed-circuit television was on display. I saw myself on TV for the first time. When I asked my father how the picture got from the camera to the screen he pointed to a cable taped to the floor. This was disturbing. How can a picture travel through a wire? I knew that cameras could take pictures on film, but how could they push pictures through wires? He told me to look more closely at the screen. I stuck my nose up to the picture tube and saw thousands of little squares with lines between them. Some were lit up white; others were dark, and others a shade of gray. As the camera scanned a small crowd of people standing nearby, lights in each little square of the screen went on or off. I tried to look at only one or two squares, but my concentration was diverted by lights flashing on and off in squares on either side. The larger images they made were distracting me. I backed up an inch or two. At a mid distance between single squares and recognizable objects I noticed there was no clear point where an object came to an end. There was a blur around its edge that was unnoticeable from farther away. If an object were really small—smaller than the squares—it might not show up at all. It might get lost somewhere in the area between one square and another. I looked up at a boy about my age in a white shirt waving his arm back and forth in front of the camera. When I looked up close at the screen I did not see the shirtsleeve; what I saw were lights coming on where it was going and off where it had been.

Standing behind me my father said, "The camera sends little electrons through the wire that tell the screen which lights to turn on."

I asked him to stand in front of the camera so I could see him on the screen. "So Dad," I said. "That's not really you, then. That's just a bunch of little electrons that turn into lights that look like you when I stand back."

"It's an *image* of me." He said.

"But it's not really you," I said. "It sure looks like your face."

"No, I'm over here."

Years later I took a survey course in physics. The professor said that a subatomic particle moving through space makes a series of dots that indicate where it was at particular points in time. But there is no way to know where it was between the dots, or if it was *anywhere* between the dots, or even if it is the *same* particle that connects the dots. It's not a question of accuracy of measurement—no matter how refined our instruments are we cannot find the particle anywhere between the dots. There is no space there. The space-time universe falls apart on this extremely small scale. The dots look like a line or a curve and we think of a little particle traveling along a path, but that is just how we connect the dots. That's just how we picture it. Larger objects are easy to locate in space and time; the problem is with the tiny ones that are smaller than the space they are in. But even larger objects do not have clear edges.

Rod and cone cells in the retina send electrical impulses through the optic nerve to the brain where they are arranged into the picture that we see. We think of the picture as a copy of what is really out there, but there is no way to know what is out there except through electrical impulses in our brains. There is no way to see beyond vision or feel beyond touch—no way to experience outside of experience. Something appears to be out there because where and when we see objects is where and when we touch them.

The quantum screen is not in space. It is not in your frontal lobe, or anywhere in your head.

It is not a screen. It is only like a screen.

It is not made of anything.

It is not made of quanta, as they are normally understood. The quantum screen is the grid pattern that quanta fall into.

A quantum is a piece of energy: the smallest and most basic piece. An object moving through space is said to have energy of motion; because energy is quantized, the object does not move perfectly smoothly through space. It jumps from one set of quanta to the next, the way an object is seen to move across a pixel screen. By *quantum screen* I mean the set of points in space and time through which the object moves. The object itself is only the pattern. The screen is the set of points in space, time, and mass that the pattern moves across as you see, hear, smell, taste, and touch it. The analogy to a pixel screen is near perfect, though a flat computer or television screen has two dimensions of space and one of time, where the quantum screen has three of space, plus time and mass. A holographic screen would improve the analogy by one space dimension, and virtual reality, by adding the mass dimension to some parts of the body, would turn the analogy (virtually) into the thing itself.

Each photon is a point of light of a particular wavelength, or color. It is a point in space-time, where a quantum is a point in space-time-mass. The mass dimension allows you to coordinate what you touch with what you see.

The photon screen is a part of the much larger quantum screen. Where the photon screen is actual photons, the quantum screen is potential photons outside your field of vision. Objects that you hear, smell, taste, and touch are at locations on the quantum screen where you will see them if you look. When you look, quanta are actualized in the form of photons. They light up. When you turn your head toward them they fill with color.

Light underlies consciousness as a whole, even when it is dark.

The quantum screen is potential perceptual consciousness as a whole, which is the same thing as space-time-mass. It is where perception happens, and where all physical objects exist. You always touch or see an object exactly where and when you hear or taste it. This gives the object an illusion of existence outside of perception; it seems to be out there waiting to be perceived. This is why we think there is an external world causing us to see, touch, or hear things.

The quantum screen does not require matter.

The existence of matter is an assumption. It is a deep assumption that underlies the Western scientific worldview, an assumption so deep that most scientists do not realize it is an assumption. If matter does exist, consciousness is a reproduction of what is really "out there," an imperfect copy of the physical world. But there is no proof that it does, and there can be no proof. There is no way to experience anything outside of experience. If you make the assumption, you have to explain the enigmas of quantum mechanics and relativity theory in some other way. A hundred years have passed and there is still no material explanation for the Heisenberg Uncertainty Principle or the relativistic increase of mass at high velocities.

The quantum screen presents a picture that is itself physical reality. Physics is in the screen, not hidden in the behavior of material objects. It is a picture, but not a picture *of anything*. On the macroscopic level the picture looks the same as the material picture. It is only at dimensional extremes that the graininess of the screen begins to show and the picture falls through the cracks. But it is only a picture.

As the picture falls through the cracks we see that it is a picture.

When we understand it as a picture we no longer look for anything between the cracks. We no longer look for matter.

Mass is what you feel. It is the heaviness of rock or the weight of your body. It is the impact of air molecules on the cells of the tympanum or of photons on the retina.

On the macroscopic level you cannot tell an object's mass by its location in space at any one time. You can tell it only by the object's behavior through a series of points in time. The more massive, the slower it gains speed when subject to a known force. It is easier, for instance, to push a basketball than a bowling ball. Acceleration is measured in terms of a second time dimension. Where constant velocity is measured in meters per second, constant acceleration is measured in meters per second *per second*. An object subject to a constant force accelerates at a constant value in this dimension, inversely proportional to its mass. If you push a basketball with the same force as a bowling ball the basketball will accelerate at a higher rate. Each second you are pushing it, it will gain more velocity than the bowling ball.

The Mass Dimension

Calling mass a dimension may be troublesome to some readers. We do not experience mass the way we do other dimensions; we are not *in* it the way we are in time or space. But there are several good reasons I understand it this way. First, the only way to measure mass is (indirectly) through a second time dimension: the per second *per second* of acceleration (or resistance to acceleration). Second, mass is inextricably intertwined with space and time in both quantum mechanics and relativity theory. It interacts with them the way they interact with each other. Mass is an equal factor with space and time in the Heisenberg Uncertainty relation and Debroglie Waves, and dilates at the same rate as space and time in Special Relativity. Third, there is no explanation for the properties of inertial or gravitational mass even on the macroscopic level. There is no mechanical understanding for how mass grabs onto space-time when a body is accelerated, or curves space-time near gravitating bodies. Fourth, there is no physical explanation for why we feel acceleration and do not feel velocity. Fifth, though we are not *in* the mass dimension, neither are we in time or space, as I have been trying to show.

Mass is traditionally accepted to be a measure of material substance—the amount of physical stuff that is in an object. But if we do away with the concept of matter (which we must in order to understand modern physics) we need no longer think of it this way. We can continue to do physics understanding mass as the fifth component of the five-dimensional nexus of physical reality: space-time-mass. Physics works just as well this way. Mass appears in four-dimensional space-time as a foreshortening into the time dimension.

The mass dimension is foreshortened within each object in four-dimensional space-time. Foreshortening occurs when there are not enough dimensions to show the fullness of an object, as when a three dimensional figure is projected onto a two dimensional surface. A photograph, for instance, shows three-dimensional objects in two dimensions. The distance dimension is foreshortened into the other two. Overall space is reduced by one dimension, but you can tell it is there by the foreshortening within each object. Similarly, the mass dimension is foreshortened in each object in four-dimensional space-time. You can tell mass is there by the behavior of each object, though it is nowhere in the space between objects.

The mass dimension as a whole affects the quantum screen as a whole when under acceleration. Acceleration is any change in the magnitude or direction of velocity—going faster, going slower, turning right or left, or going over a pothole. You cannot feel velocity (the first time dimension) in your body, but you always feel the second time dimension. Everything you feel is an acceleration of some sort. Usually, your body does not move as a whole when you touch something because you balance the acceleration with an equal and opposite acceleration somewhere else in your body. When somebody hands you a book your feet automatically push back on the floor in the opposite direction. But when your body accelerates as a whole—speeding up, putting on the brakes, going around a curve or over a bump—you feel a uniform kinesthetic sensation in exact proportion to the per second *per second* of your motion. You feel the kalapas built into the quantum screen.

Science has never explained why you feel acceleration but not velocity. You can be traveling constantly at a hundred, or a thousand, or a million miles per hour and not feel a thing, but you feel the slightest *change* in velocity. How does your body know to not feel the "per second" but to feel the "per second *per second?*" And how, for that matter, do physical objects cling to space to resist acceleration? The phenomenon of mass and acceleration is commonplace—it does not involve quantum physics—yet there is no other explanation for it in classical or modern physics.

Mass, traditionally thought to be a measure of material substance in physical objects, is better understood as the fifth dimension of the quantum screen.

Quanta are arranged as five-dimensional points on the screen. Physical objects are ranges of contiguous points on the screen. At dimensional extremes the screen itself becomes distorted, as if you were looking at it at an angle. Points become closer in space and farther apart in time and mass. At nearly the speed of light you are looking at the screen nearly edge-on.

All dimensions are interchangeable. An observer traveling at a uniform velocity "through" space interchanges one space dimension with time. An observer accelerating in a straight line interchanges one space dimension with mass, and feels the mass dimension as a whole throughout his body. An observer accelerating around a curve interchanges two space dimensions with time and mass.

If the speed of light were infinite and the Planck constant zero there would be no way to know of the quantum screen. There would be no uncertainty principles, no cosmic speed limits, no dimensional dilations at high velocities, and no way to glimpse the hidden structure of the universe. It is only where the screen falls apart that we see that it is there, only where its edges peel that we glimpse the void beneath. The edges begin to peel at dimensional extremes: the very small, the very fast, the very massive, and also, at the very distant.

Let us consider the very distant. If you look far enough in any direction you should see the end of the quantum screen.

If you look far enough into space you should see the beginning of time.

You don't see it because there are too many intervening objects and your eyes are not good enough. You need a clear, dark night, and you need a big telescope—a really big one, and one that can receive all wavelengths of electromagnetic radiation, not just visible light. The reason is that the universe is expanding in all directions, and the farther away an object is, the faster away it is moving. As objects move away, wavelengths of light coming from them are stretched out in space. The faster they move, the longer their wavelengths become. This makes them appear lower on the frequency spectrum and changes their color. Visible light becomes red-shifted, or moved toward the longer-wavelength red end of the light spectrum. More distant objects are moved out of the visible light spectrum entirely and into infrared or radio wavelengths. The most distant objects become so distorted that you cannot see them at all, even with the largest telescope. They no longer consist of visible light.

The most distant "object" on the quantum screen is the *cosmic background radiation*, composed of very low frequency (long wavelength) photons "left over" from just after the big bang. If you pick them up on a radio telescope you are looking at something that happened several billons years ago, but you are looking at it *as it happens*. You are looking back that far in time.

At extreme distances space is expanding away from you so fast that velocities approach the speed of light. According to the special theory of relativity, time slows under these conditions and mass increases. At the speed of light itself—at the edge of the universe—mass becomes infinite. According to the general theory of relativity, mass curves space-time: infinite mass curves space-time infinitely. The quantum screen turns back on itself to a single point. The extreme edge of the universe is a single point—the singularity from which the big bang exploded and the universe emerged. It is a point in all directions around you. The entire quantum screen is inside of it.

You can't see the singularity because there were no photons then. The screen falls apart at that point. You have to wait until photons emerged a few hundred thousand years later and became the background radiation. The background radiation is everywhere in all directions, but despite its name, it is not really the background. It is the first *thing*, the first physical object against the background of emerging space, time, and mass. The real background is the quantum screen, a little way behind, just beginning to sort itself out.

The meditation practice taught by the Buddha many hundreds

of years ago, when expanded to include vision, reveals a connection between minute body sensations and quanta. It is a connection that cannot be understood objectively. It cannot be proven scientifically. The reason for this would seem on the surface to be that it is only your subjective experience; it is in your body or only in your mind. There is no way to measure or verify it, and no way to convey it to other people other than through suggestion. Science does not accept suggestion, and should not. Science requires measurement and verification. You have to be able to say how long something is, how fast it is moving, and how much mass it has. In other words, you have to know its dimensional components. It has to be in a space-time context. It has to be large enough, slow enough, and small enough to be on the quantum screen.

Video screens, retinas, brains, and optic nerves are all objective phenomena in space and time. Even cells are large enough to be objects. Because these are all objects on the quantum screen there is a limit to how far they can be used to understand the screen itself. If you picture a retinal cell hit by a photon that has flown across the room, converting it into an electron that travels through the optic nerve to be arranged in the brain with other electrical impulses, you are using a series of dimensional concepts to get at a reality that is more fundamental than dimensions. You are using dimensions to see what dimensions are. It may be helpful in relating something known about anatomy to something known about light, but the concepts break down when the space that is created must already exist as the photons are flying across the room. You can't use space, or any sort of spatial relation, to get at what space itself is. Quantum physicists have been struggling with this problem since the days of Max Planck. You cannot use dimensional concepts of everyday life to understand what happens when dimensions break down. You have to find some other way of thinking about it. Or, you have to look beyond thinking to what is there.

Objective knowledge of cellular structure is useful in understanding the structure of the screen, but the screen, because it is experience, cannot be understood ultimately as a product of cellular structure. The dimensions within which you understand the size, shape, and bio-chemical functioning of cells are themselves products of the screen.

The quantum screen explains modern physics where other models fail because it is not an object. It is not in space and does not exist without experience. For the same reason, it is not science. It includes science but is not science.

The screen has no physical relation to anything else. It is what is there.

With practice and good concentration you will see, hear, taste, smell, and feel quanta exactly as they are. You will see them directly, in the world, the way you see kalapas in the body. You will hear the orchestrated voice of every cell. You will experience the wholeness of each realm and the cross referencing of each dimension in every object. You will taste and smell a quantum pattern on the screen and feel the body as a shape in five dimensions. The diversity of objects in your field of vision will melt into a single thing. The solidity of the material world will dissolve into quanta, as gross sensations dissolve into kalapas.

There will be a moment of singularity that is not you, a moment not in time. This is bhanga—bhanga with space—an external bhanga without the externality. This is *quantum bhanga*.

Quantum bhanga, like *quantum meditation*, sounds a little too good. Some will say it when they mean much less, and I hesitate to use the term. But it is exactly what I mean. I hesitate because it is no more than a concept when not directly experienced, by you. The term is not what it is. It sounds like some altered state of being, some blissful, esoteric, and desirable goal—some end worth striving towards. It could be the consummate experience for those seeking a unified vision of the physical world. It could be what the physical world is, within a larger vision of consciousness as a whole. It could serve in this manner. It could be an understanding of the reality we want to know in our lives, and that would be fine. It could explain the enigmas of quantum mechanics and relativity theory, and it may be why I have written this book.

But it is a concept. *Quantum bhanga* is a thought and not an experience until it is experienced. It is transcendence of the material world and of all divisions created by the self—a wholeness not in space, a stillness not in time.

You can think about it but it will not be what it is until you experience it.

This, though you, as you, cannot experience it.

DIMENSIONAL
INTERCHANGE

When you slip back in to space and time, and into mass, the wholeness divides. The moment is gone. Experience becomes memory, and memory turns to concept. You cannot maintain quantum bhanga through time because it is not in time.

But there is a sort of walking meditation that interchanges time with space and mass. It goes on all the time. You do it because that is what doing is. When you are in time you constantly interchange one dimension for another, rotating the axes of space, time, and mass on the quantum screen.

Quantum bhanga is the unity of all experience. It is everything that is there, but it is static. It is complete passivity. You are not there and it does not need you—but neither does it explain your apparent existence when you are there.

Dimensional interchange is a complexity that explains why you seem to be moving around doing things in space. It is a complexity that allows for the unity of existence when you are walking down the street.

You feel you are in space because you move in space. But spatial relations are relative; your motion through space can be understood as the motion of space through you. Moving westward down the street can be understood as the street—and the rest of the universe—moving eastward.

Constant velocity is interchange of one space and one time dimension. Time becomes space: the passage of time becomes the passage of space, at a steady rate. If the rate is fast, you get more space for each unit of time. Instead of seeing yourself walking down the hallway, you see yourself "scrolling" through space at a constant rate, as doorways, ceiling lights, and floor tiles pass by on all sides.

Another way to understand this is as a rotation of the axes of space-time. Picture yourself standing on a giant piece of graph paper that fills the floor of a large room. Each square on the paper is a square meter. You notice a chair ten squares in front of you and three squares to the left. You decide to call the directional lines running forward and backward in front of you the y *axis* and the lines running left and right the x *axis*, so the chair, relative to you, is at x = - 3, y = 10. If you are on the floor and the chair is hanging by a rope 9 meters above the floor, you include the up-down dimension as the z axis, and the chair is at x = - 3, y =10, z = 9. But this is no more than an arbitrary coordinate system that you have invented for your convenience. If you preferred, you could rotate the axes horizontally 90 degrees to the right and say that the chair was at x = 10, y = 3, z = 9, or rotate them vertically and say that it was at x = -3, y =-9, and z = 10. You could complicate things more by rotating the axes only 5 degrees, or 10. Either way, it would be looking at the same thing in a different way. The chair would still be where it is. But now let us say that you have heard about the theory of special relativity, and want to include time, the t *axis*, to your relation to the chair. This does not change anything if you are standing still: all of the spatial dimensions remain the same whether you are at t = 0, t = 100, or anything else. But if you rotate the t axis slightly into the y-axis, at say the rate of 1 second = one meter, you will find your spatial relation to the chair changing in time. After ten seconds it will be at x = - 3, y = 0, z = 9. This will look like you are moving through space closer to the chair.

Dimensional interchange is key to understanding the relation of the body and space, but it is also a clear demonstration that the apparent motion of the body through space does not mean that consciousness is *in* space or time. The simple interchange of space and time described above is understood to be a rotation of axes within perceptual consciousness as a whole, as opposed to within an external universe. The more complicated dimensional interchanges described below can be understood only as within consciousness.

With the t axis tilted slightly into the y-axis you see yourself moving slowly across the floor toward the chair. But how did you get moving in the first place? More importantly, how will you stop? With the axes rotated you will keep moving past the chair and run into the wall. How do you *change* your motion, in rate or direction?

For this you need to interchange another dimension: the *mass* dimension. Whenever you begin to move or stop moving, increase or decrease your speed, or change your direction, you enter the second time dimension: you *accelerate*. Your motion is no longer constant velocity and can no longer be described in terms of meters per second (space per time). It is describable now only in meters per second *per second*. This is the dimensional combination that describes the inertial behavior of massive bodies in space-time.

To go from standing still on the floor to moving at one meter per second you have to accelerate—you have to rotate space-time axes into *mass*. You become non-zero in the mass dimension only as the time axis is in the process of rotating; mass returns to zero once you are through accelerating and settle in at a constant velocity. Perhaps you accelerate only for the first second, go back to zero acceleration, and continue across the floor at a steady one meter per second. But then, as you sail past the chair and see the wall coming toward you, you decelerate (a form of acceleration) by rotating into mass again, perhaps only slightly, to stop or to change direction. You can tell you are non-zero in the mass dimension by the change in the second time dimension of your motion and by the uniform sensation you feel throughout the tactile realm. The entire mass dimension is activated at exactly the time and rate that the universe appears to accelerate past you in the opposite direction.

The mass dimension, you will notice, is hardly ever at zero. You cannot feel constant velocity, no matter how fast you are moving, but you know you are moving by the slight accelerations you feel—you step on the brake, round a bend, or go over a bump. As you walk or drive down the street you feel small accelerations in the bounce of your step, the vibration of the wheels, or the unevenness of the pavement. You may also notice that the ground you are standing on produces a constant acceleration upward against the earth's gravitational field. You feel this acceleration all the time, and when you feel it you see all the objects around you accelerating in the opposite direction until the floor supports them, too (or what remains of them). The only time you do not feel this *g* force is when you are freefalling off a tall building or drifting listlessly through interstellar space. Only then is mass zero, and you feel nothing.

Interchange of the mass dimension is how you change motion, and it is through changes in motion that you do all the things that you do. But how do you change the *right* motions to go over to the chair, to avoid crashing into the wall, or to do anything else in an orderly fashion? How do you know when and how to make the orderly accelerations that get you out the door and on the bus, that move your fingers across the keyboard, or that move your diaphragm and vocal chords to say something sensible? All of these are dimensional interchanges involving some part of the body, and therefore, the mass dimension. They are all forms of *doing*; you always feel the body when you do something.

Without *order* you would feel the mass dimension, but you would still crash into the wall, write gibberish on the computer, and make nothing but noise with your vocal chords. You would be no more than an inanimate object and unable to do anything sensible. Orderly motion is what distinguishes you from an inanimate object. Because you are an observer, and animate, you are able to interchange space, time, and mass dimensions, and also interchange mass with an orderly arrangement that you have thought of. The thinking that becomes doing, or the *practical realm of consciousness*, is not perceptual—objects of doable thought are not in space, time, or mass—but it becomes perceptual *as it is done*. You can see what you are thinking of doing after you do it.

Inanimate objects accelerate *uniformly*, but you, like other observers, are capable of orderly *non-uniform* acceleration. The order that you create by doing things shows up as yet another dimension foreshortened on the quantum screen.

Unlike other dimensions, order is not quantifiable. You cannot measure it in meters, seconds, and grams. Its potential is foreshortened in space-time, as is that of mass, and does not show up in the structure of the screen—it does not appear as empty space between objects. It reveals itself instead in the behavior of each object that has the potential for observational consciousness. It shows up as the orderly acceleration of observers. It is how you can tell they are alive.

The Order Dimension

Through order we become aware of observational consciousness, that is, of other living things. We can tell they are alive by the way they move. Order acts like a dimension in some ways. Where inanimate objects accelerate uniformly, observers change the direction or rate of their acceleration as if in another dimension. But unlike space, time, or mass, order is not quantified. (This may have to do its interchangeability with only one other dimension, mass.)

The potential for observational consciousness is *dis*order, or the tendency toward entropy throughout the universe. Order is actual observational information within this potential. Living things are orderly against the background of general disorder.

As the quantum screen is only five-dimensional, order cannot be perceived directly. It is foreshortened on the screen within each object that shows potential observational consciousness.

Where time, mass, and space correspond to perceptual realms of consciousness, *order* corresponds to the observational realm.

The place in your mind that thinks of things that can be done is the also the place where you put the information you hear from other people. Doable thought (the practical realm) and observational consciousness appear on what I call the *image screen*. The image screen is an extrapolation of the quantum screen, without the quanta. It is an infinitude of possibilities beyond the actual. But it is not all possibilities, only those that coordinate with the dimensions of the quantum screen—only those that obey the laws of physics. It is not perception. It is not as clear or precise as the quantum screen—it has the structure without the high definition of light. Yet it is one dimension greater than the quantum screen in that it shows not just what you see, but what you *might* see. Where the quantum screen is five dimensional the image screen is six dimensional. Its substance is of ideas, words, numbers, and images, etc., but not all ideas, words, and images—only those that can be shaped into dimensions: those that can be done. They have to be potentially perceivable. If you are listening to a description of something you are not seeing, or thinking of something you plan to do,

you experience dimensional patterns on this screen. Its coordination with the quantum screen allows you to experience the
same pattern perceptually if you go look in the right place, or
actually do something you have been thinking of doing.

> For practical thought, the image screen is a tool of self.
> For science, it is the universe of all observers.

It is here that you experience me describing the furniture in
the next room, a busy street corner in a distant city, or the
crater pattern on the far side of the moon. And it is here that
you conceive of what to do next.

What you do, of course, does not always turn out the way you
thought it would. The quantum screen does not always show
what you saw on the image screen before you started. In fact, it
never does, because you cannot locate images perfectly. Your
knowledge is never complete and the screen you are working
with is never as clear as actual perception. You cannot know
where everything is, how it is moving, and how massive it is.
You cannot make out the details of conceived images. The images or parts of images that survive the interchange of practical
thought with mass (as you make the effort of doing) are those
that coordinate perfectly with space and time. The undoable is
weeded out. So you try again, this time with a better picture on
the image screen. The more accurately you locate images the
more successfully you do things.

If, for some reason, you think you could make better use of the chair somewhere other than at x = - 3, y = 10, z = 9, you locate it on the image screen where you would like it to be. This is not as easy as it sounds. You have to be able to do what you are thinking of doing. Imagining your body floating up nine feet off the ground and moving the chair from z = 9 to, say, z = 14 would not coordinate with the quantum screen. Doable thought is restricted to the physically practical. So, instead of flying through space, you decide to move the chair to x = -4, y = 10, z = 0, where you and the chair will remain supported by the floor. This is doable. You see the chair where you want it on the image screen. Then you interchange the right dimensions to move your body to the tool shed and get a ladder. You set the ladder at x = -3, y = 10, z = 0, climb up and cut the cord holding the chair at z = 9, bring the chair to the floor, and set it down at x = - 4. The thought process may be so rapid as to be barely conscious. It may happen so quickly that you do not fully picture all the intermediate steps. But you can do it only because you thought of it.

Seeing the chair in its new position is a manifestation in space-time-mass of your thought experience in the practical realm. Once you coordinate your body with what you think, and interchange all the right dimensions, the new arrangement passes from the image to the quantum screen. You can see it where you only thought it before.

Dimensional interchange is stillness in motion.

It is the stillness of doing, without the stillness of sitting still.
It is watching self, without self.

Self is the axis of rotation.

SELF

Self is a focus of order.

Self stretches beyond space and time, beyond mass, all the way to thought. It reaches across the dimensional realms, into imagination and doable thought, searching for order and bringing its version of order to physical reality. Your self does things, with the help of your body. It brings order into physical reality by coordinating and interchanging practical thought with the tactile realm.

Other selves are other people, animals, church groups, armies, nations, sewing circles, and football teams. Each is order, each with its own focus. You may fit in.

The self is always fussing, always dissatisfied with the way things are. It wants things to change. It wants you to improve things. It wants you to move your leg when it hurts. If you are unhappy with the way things are, your self will help you make them better. If you are happy with the way things are, your self will make you unhappy.

It is the ache.

Self is most difficult to see from the outside because it goes to near the beginning of thought.

If you would see self, try to do nothing.

The struggle in meditation is with self. You must engage in the struggle with the fullness of mind, heart, and body, while watching from the sidelines. You must lose, and not give up. If you win, self wins.

Being does not need you.

But saying that self is a focus of order does not tell you how one consciousness is connected to another. It does not tell you how "your" consciousness is connected to "mine" or someone else's. It does not say how what you see in the body as a subtle sensation becomes a quantum particle in a physics laboratory or how what you perceive is connected to what a scientist perceives through his telescopes and microscopes and particle accelerators.

   Telescopes and microscopes and scientific instruments of all kinds extend perceptual consciousness, but they have no bearing on the relation of one consciousness to another. They do not change the structure of consciousness. What you or I or anyone else sees through an instrument may be an amplification of what we would see without it, but it remains seeing. Perception remains perception. Even if what you would hear is converted through instrumentation into a visual representation, it is shifting auditory to visual perception, without changing anything about the relations between them. The question of the relation between science and perceptual experience is not, then, one of instrumentation. The question is, "How does what someone else sees become what you see?"

And there is no answer.

What someone else sees has no physical relation to what you see. You have no experience of what someone else sees, only of what they *say* they see. From the standpoint of perceptual consciousness "someone else's" perceptual experience does not exist. You will never know what anyone else actually experiences because it is not there. It does not exist.

What does exist is a structural relation between what you see and what you hear someone else saying he sees. It is a relation between perceptual and observational consciousness. Self is not involved. (If your self or his self tells a lie, the structure of observational consciousness is damaged or destroyed.) "Your" consciousness and that of the scientist in the laboratory have no separate identity. Your experience is of kalapas in your body that become quanta in space-time, and of the scientist's *description* of a quantum. That is all you actually experience. If his description accords with what you or anyone else would see in his place, observational consciousness is created. Your self and his self have nothing to do with it. It is a question not of one consciousness and another, but of one realm of consciousness and another.

This is how the observational realm appears from the standpoint of perceptual consciousness. Observation is reducible to the perceptual experience of hearing words or seeing print. But it is also a wholeness unto itself. The order implicit in information you perceive becomes a reality over and above perception alone. As organic consciousness is more than the sum of cellular experience, so observational consciousness is more than the sum of perceptual experience. As touch to individual cells becomes vision to an organism, so the experience of individual observers becomes science. As more of what you experience comes through the observational realm, science becomes a level of being that transcends perceptual experience.

Science, as a structure of consciousness, is not a focus of order. It is not a self. Experience gained through science may be used by a self—often a collective self—to create a practical order that can be done: space travel, polio vaccines, and hydrogen bombs are all orderly arrangements from some point of view. But they are applications of science, not science. Pure science is a transcendence of self.

Science is watching.

The highest form of humanity is the self of all people, a self that barely exists. It has not had to exist because we do very little as a whole. The collective human selves that now do things are mostly against one another. Each sees the order of its own interest. We are many conflicting selves, and not whole.

In relation to the wholeness of the forests, oceans, earth, and climate, this used to be fine because there was nothing we had to do. They were there no matter what we did.

Now we have to do something to keep them, and there is only the wholeness we are not.

We have no means to act.

To do, we will become.

YOU

I address this to you.

You who sees, who hears, who feels. There is no other.

You who thinks.

You.

Look at what is really there.

You see seeing; you see hearing; you see feeling.

You see thinking.

Do you see me seeing?

Do you see anyone else hearing or thinking?

Look at what you actually see.

You hear me say that I see. You see the pain in a person's face. You smell what everyone else says they smell.

You watch me hand you the morning paper.

It all checks out. Where you see the keys on the table is right where you heard me say they were. Everybody says the food is good. We all jump when you hear the sound. The appointment is at the time and place you were told it would be.

What you hear people say and what you see them do is coordinated with what you perceive. There is order in the sounds you hear coming from my mouth, and from the things my arms and legs do, that is dimensionally aligned with what you see and hear. It is coordinated with the perceptual realms the way they are coordinated with each other.

Hearing or reading a language that you understand is more than ordinary hearing or seeing. It is the hearing and seeing of symbols that stand for more than they are intrinsically. They are reducible to hearing and seeing—to guttural sounds and blotches of ink on paper—but the order in which they appear creates a separate realm of consciousness. Observational consciousness comes to you through the medium of perceptual consciousness but is much more than perception. It is built of perceptual bricks as vision is built of tactile bricks, but it is an architecture of its own, an architecture of language.

Physical objects move towards disorder over time. Things fall apart; rocks roll down hills; stars collide with planets; gates fall off hinges. Physical objects that are observers, on the other hand, create order. They move, eat, grow, communicate, and build things that look orderly. They manifest an additional dimension. They cannot do otherwise—that is how to tell them from non-observers.

The order they create comes with perspective. Each observer is a focus of order. What looks orderly to me might not look orderly to you. What looks orderly to a termite might not look orderly to a human being. An observer—a living being—curves energy to its own perspective. That makes it alive. Life is order.

Where gravitation is a curvature of space-time into mass, life is a curvature of space-time-mass into order. Life is a curvature of energy.

Doing things is an interchange of mass with order—you move your body in a way that bumps into things and rearranges them into a pattern you thought of.

You don't see me seeing things. There is no quantum screen in me. What you see is me doing things in an orderly manner, foreshortened on the screen. I talk in a way that is specially ordered so that you may see what I speak of for yourself, if you look. You perceive me making sounds and symbols on paper that create the observational realm of consciousness. It is not in me or in you.

You rarely see distant galaxies for yourself. But you know that you would see what observers say they see if you had the chance. You take their word for it most of the time—you do not bother to go to their laboratories to check—but you believe what they tell you because you believe you would see it, too.

Observational objects—things you only hear about—are not on the quantum screen. But observational *information* is—you hear it. The reduction of observational information to perceptual consciousness means that there is no separate consciousness in me or in any of the other people or plants or animals you see around you. There are only levels of consciousness that you experience in separate realms.

Perceptual consciousness—what you think of as your perspective in space—is a level of consciousness. It is a level between cellular and observational consciousness.

Consciousness is one.

It is one that is not a number.

You, here now—you, in the room with tables and chairs, with people, you, who see yourself as one of many, your being contained in self: what do you really see?

You, who are a bunching together of your cells: what are you? Are you in one cell, or between them?

The coat hanging on the rack, the bowl on the table: you

think they are in the space you are in. It looks that way. You touch the counter, hear the clock ticking, someone calls from the next room. You didn't make this all up.

The people who brought you telephones and steam engines, atomic energy and artificial fertilizers—the people who know what matter is made of and how it works—are now saying that someone has to be watching it for it to be. It is all consciousness. What you are looking at is consciousness. The little things that big things are made of do not exist on their own. They are not out there. The space and time of the universe you thought you were in are a way of assembling the information you taste and hear. It looks like you are in space and consciousness is in you, but that is only how it looks.

You go deeper than space, deeper than time. But then you stop.
   Thought is deeper still.
   Self is thought that curves thought.

Who, then, are you?
   You are physically unique. Only you see, think, taste, feel.
   This is dangerous to know, if you are still you.
   This is something you should never know.

This is why morality precedes meditation in the Buddhist religion. It is why compassion overpowers the logic of self-interest. If you do not know that consciousness is not self you may see only you when you look at what is. You may think that you are what is.

Look at the pain in your leg. Look at the thought of moving your leg. Look at the thought of *mine*.

It is the ache of all beings everywhere. You can see the whole thing.

Quanta are the limit of objectivity, kalapas of subjectivity.

Neither is either.

You are the limit of language.

APPEARANCE

$\int$uch is the distraction of open eyes. You need none of it.

Seeing is as before.

The difference is understanding (for what it is worth): no external world, no you.

This, from the Buddha, and from the quantum.

Physics makes better sense if we do not assume what is not there. Material substance in physical objects and consciousness in observers—like ether and a flat earth—look like they are there but we do not need them and they are in the way. All we need is the dimensional coordination of separate realms of consciousness. You see something where you touch it; you touch something where someone else says it is.

In the middle latitudes of space-time-mass—where we live our everyday lives—things make sense with or without the common sense assumptions. It looks as if matter could be there and as if the earth could be flat. At dimensional extremes it does not. The oceans curve and matter no longer fits in the box.

Since 1900 the mind has traveled to the very distant, the very small, the very fast, and the very massive. It has gone beyond the middle latitudes. It has touched the limits of physical reality and cannot shrink back to where it was. We can no longer think of a world independent of consciousness. If we do, that is all we are doing.

The quantum screen is a model: an analogy, a way of thinking. It is not really there. What is there is what the words stand for.

We have seen past it, around it, through and behind it. We have seen its coarse surface and frayed edges. It is flawed, warped, imperfect, and elegant. It is the architecture of multi-cellular consciousness.

To know water
Drink it.

To know appearance
Open your eyes.

Draw water uncontained.

Transcend self.

Teach.

ABOUT THE AUTHOR

Samuel Avery holds degrees in religion and history. He has practiced meditation daily for forty-one years.

Samuel has written a series of articles and books on the relationship of physics and consciousness, including the books *Transcendence of the Western Mind* and *The Dimensional Structure of Consciousness*. He installs photovoltaic systems for a living and lives with his wife on a small farm on the banks of the Nolin River in Hart County, Kentucky.

Sentient Publications, LLC publishes books on cultural creativity, experimental education, transformative spirituality, holistic health, new science, ecology, and other topics, approached from an integral viewpoint. Our authors are intensely interested in exploring the nature of life from fresh perspectives, addressing life's great questions, and fostering the full expression of the human potential. Sentient Publications' books arise from the spirit of inquiry and the richness of the inherent dialogue between writer and reader.

Our Culture Tools series is designed to give social catalyzers and cultural entrepreneurs the essential information, technology, and inspiration to forge a sustainable, creative, and compassionate world.

We are very interested in hearing from our readers. To direct suggestions or comments to us, or to be added to our mailing list, please contact:

SENTIENT PUBLICATIONS, LLC
1113 Spruce Street
Boulder, CO 80302
303-443-2188
contact@sentientpublications.com
www.sentientpublications.com